EAST AFRICAN WEEDS
AND THEIR CONTROL

G. W. IVENS

OXFORD UNIVERSITY PRESS
Nairobi

Oxford University Press

OXFORD GLASGOW NEW YORK
PETALING JAYA SINGAPORE HONG KONG TOKYO
DELHI BOMBAY CALCUTTA MADRAS KARACHI
NAIROBI DAR ES SALAAM CAPE TOWN

and associates in
BERLIN IBADAN

ISBN 0 19 572530 1

© *Oxford University Press, 1967*

OXFORD is a trademark of Oxford University Press

Reprinted with corrections, 1968, 1971, 1975

2nd edition, 1989

Published by Oxford University Press, Eastern Africa, Science House,
Monrovia Street, P.O. Box 72532, Nairobi, Kenya and printed by
English Press Ltd., P.O. Box 30127, Enterprise Road, Nairobi.

Foreword

by E. W. RUSSELL, C.M.G.,

lately Director of the East African Agricultural & Forestry Research Organization.

It gives me great pleasure to introduce this book to both the agricultural community of East Africa and to everyone else who takes a practical interest in raising the agricultural productivity of this very important part of Africa. Weeds cause an enormous loss of produce every year, and no improved system of agriculture can be developed unless it ensures that the weed population in the land can be kept at a low level efficiently and economically.

The need for this handbook was realized at a Conference held under my Chairmanship at the headquarters of the East African Agriculture and Forestry Research Organization devoted to the role of the modern range of herbicides in the cultivation of the principal East African cash crops. These herbicides are selective, in that when properly used, they only kill certain species of plants and do not affect others. They can therefore be selected to control a certain range of weeds, but to have no harmful effect on the crop. A suitable herbicide can only be chosen, however, if the weeds can be correctly identified, and if it is known whether they are susceptible or resistant to the various chemicals available.

At the time of the Conference there was no handbook available which described or pictured the weeds of East Africa and which would enable the farmer or the agricultural extension and advisory staff to put the correct name to the weeds they encountered. The author consented to be responsible for preparing such a book. He was, at the time, the botanist at the Tropical Pesticides Research Institute at Arusha, and was engaged in evaluating under East African conditions the new herbicides being produced in Europe and America. He therefore has first-hand knowledge, both of the more important weeds in the principal crops of East Africa and of the herbicides likely to be of value for their control.

In completing this work Dr Ivens has been fortunate in receiving the help of the advisory and research staffs not only of the Government departments of Agriculture and his successors at the Tropical Pesticides

Research Institute, but also of the various commercial firms concerned with the development of herbicides in East Africa. He has been particularly fortunate in obtaining the permission of the now disbanded Department of Research and Specialist Services of the Federation of Rhodesia and Nyasaland to make use of a number of the drawings included in Dr H. Wild's book on Rhodesian weeds and in being able to persuade Mr G. R. Cunningham van Someren to make drawings of the remainder of the plants.

It is my hope and the hope of all who have been in any way concerned with the publication of this book that it will facilitate the identification of the weeds of East Africa and that the information given on susceptibility to herbicides will facilitate their control.

Contents

Preface

The first steps towards compiling this book were taken in 1958, when the author was botanist at the Colonial Pesticides Research Unit (later to become the Tropical Pesticides Research Institute) near Arusha, Tanzania. In this year a questionnaire designed to obtain information on weed distribution in East Africa was circulated to agricultural officers in all parts of Kenya, Tanzania and Uganda. A total of 20 replies was received (12 from Kenya, seven from Tanzania and one from the Western Province of Uganda) and this information, together with personal observations, information from Kew and from the various relevant parts of the Flora of Tropical East Africa provides the basis for the notes on distribution which follow.

It was intended that this book dealing with the identification and control of weeds should be accompanied by a second publication dealing with the possibilities for chemical weed control in East African crops and this aspect of the subject was competently covered in 1963 by Hocombe and Yates' *Guide to Chemical Weed Control in East African Crops*, to which the present work is complementary. The author's return to the UK and the arrangement of financial support for publication have delayed the book's appearance. The original text was completed in 1963, but has recently been extensively revised to take into account the rapid developments that have taken place in chemical weed control and it is hoped that the information given is reasonably up to date.

In an area with such a rich and varied flora as East Africa it has not always been easy to decide which species to include and which to omit. It was originally intended to limit the number to 100, but it soon became obvious that this would mean the exclusion of many important weeds and would provide no opportunity of dealing adequately with the many groups of related species which are often so difficult to distinguish. By widening the scope of the book to include 100 species or groups of related species it became possible to include a representative selection of water weeds and woody weeds in addition to the common arable weeds and the number eventually grew to 110. The total number of individual species described is 220.

The book was planned to include a line drawing for each species or group of species to facilitate identification and Mr G. R. Cunningham van Someren undertook the major work of preparing the drawings. In order to lighten this task an approach was made to the Department of Research and Specialist Services in Salisbury for permission to use some of the illustrations from Dr H. Wild's book, *Common Rhodesian Weeds.* The plates were generously made available and 44 of the weeds common to Rhodesia and East Africa are illustrated from this source.

The botanical names used are as up to date as possible and have all been checked at Kew within the last year or two. Inevitably some of the names will have been changed since then and any inaccuracies should be attributed to the author rather than to Kew. English common names have been given where they exist. The value of the book would doubtless have been increased if it had also been possible to list African names, but this was precluded by the author's departure from East Africa in 1959.

The wording of the descriptions has been kept as simple and non-technical as possible and a glossary is included to explain the few botanical terms that have been used. It also covers those terms which relate to the use of herbicides.

Information on the susceptibility of the various weeds to herbicides comes partly from personal experience, partly from the experience of others in East Africa (especially that reported in the Proceedings of the African Weed Control Conference, 1958) and partly from other literature reported in Weed Abstracts. The chemicals on which most information is given are MCPA and 2, 4-D as the response to these materials varies greatly with different species, i.e. they are highly selective. Soil-applied herbicides, such as the triazines and substituted ureas, are generally less selective and will kill the majority of plants as germinating seedlings, though a number of large-seeded plants tend to be more resistant. They have relatively little effect, however, on larger, well established plants or perennial weeds, so that they can be used selectively in many woody or perennial crops. Information on these chemicals is given where available and occasional reference is also made to the contact herbicides paraquat and diquat. These chemicals are relatively non-selective and kill most types of foliage with which they come in contact. In crops, therefore, they must be applied as directed sprays and it can be assumed that the majority of annual weeds are susceptible to their effects (diquat is less effective against grasses). Reference to paraquat and diquat is thus only made where special uses exist or where resistance to their action has been found.

It is hoped that the publication of this book will not only facilitate the identification of weeds in East Africa, but also make it easier to control them. Doubtless a number of the chemicals mentioned will eventually be superseded by more effective materials, but, although the chemical information may go out of date, knowledge of the true identity of a weed will always be of value in the adoption of a herbicide technique.

Among the various bodies who have supported the preparation of this book especial thanks are due to the Ministry of Overseas Development whose generous grant will enable it to be sold at a low enough price for wide distribution to be possible. The financial support of the following firms is also gratefully acknowledged: Shell Chemical Co. of Eastern Africa Ltd., Fisons (East Africa) Ltd., R. O. Hamilton Ltd., Twiga Chemical Industries Ltd., Murphy Chemicals (East Africa) Ltd., as are the contributions from the East African Common Services Organization, and from Egerton College, Njoro.

On a less material plane my thanks are due to Professor E. W. Russell for his continued interest and support, to Dr H. Wild and Dr G. R. Bates of the former Federal Ministry of Agriculture for permission to use illustrations from Dr Wild's book, to Dr B. Verdcourt and his colleagues in the East African Herbarium, Nairobi, for identifying most of the plants described and to Mr Brenan and his colleagues at Kew who have gone to great trouble to answer my numerous questions about some of the more awkward plant genera.

Oxford, 1965   *G. W. Ivens*

Author's Note to the Second Revised Edition

Since the last revision of this book there have been many new herbicide developments, with especially notable advances in chemicals effective against grass and sedge weeds. At the same time, certain chemicals have been withdrawn, notably 2, 4, 5-T, the standard material for controlling woody weeds. The information on control methods has, therefore, needed extensive up-dating.

The weeds themselves have changed less, but a few new species have appeared, including *Parthenium hysterophorus* with a major pest potential, and there have also been various name changes to record. In preparing this revision the opportunity has been taken of adding a selection of the weeds important in rice (though a thorough treatment of rice weeds would require a separate book) together with drawings of the seedling stages of 47 of the most significant weeds of arable land.

The number of illustrations has been increased to 131 and I am grateful to the following for permitting the reproduction of published drawings: The Director, Royal Botanic Gardens, Kew, for Figures 20, 25, 27, 84 and 89 from the *Flora of Tropical East Africa;* The Kenya Agricultural Research Institute, Nairobi, for Figures 8, 9, 10, 11 and 12 from P. J. Terry's 'Sedge Weeds of East Africa: Identification' in the *East African Agriculture and Forestry Journal;* The Government Printer, Kampala, for Figure 19 from K. W. Harker's *Illustrated Guide to the Grasses of Uganda.* Figures 1, 4, 31, 67, 80, 86 and 97 were drawn by the author from specimens provided by the East African Herbarium, and Figure 74 from a photograph in the *Queensland Agricultural Journal* and material supplied by the Director, Botany Branch, Department of Primary Industries, Queensland, Australia. The assistance in New Zealand of the Photographic Unit in preparing the new illustrations is gratefully acknowledged.

I am particularly grateful to Miss C.H.S. Kabuye, Botanist in Charge, East African Herbarium, Nairobi, for providing information on and specimens of weeds; to the chemical companies for information about the available herbicides; and to Dr P. G. Otieno, Senior Agriculturist Chemist, Nairobi Agricultural Laboratories, Nairobi, for his valuable assistance.

Oxford 1987

G.W. Ivens

Classification of Herbicides

A. *Inorganic chemicals*
   sodium chlorate
   dinoseb (DNBP) 2,4-dinitro-6-sec. butylphenol

B. *Organic arsenicals*
   DSMA disodium methylarsonate
   MSMA monosodium methylarsonate

C. *Chlorinated aliphatic acids*
   dalapon 2,2-dichloropropionic acid (sodium salt)
   TCA trichloroacetic acid (sodium salt)

D. *Quaternary ammonium salts*
   diquat 1,1'-ethylene-2,2'-bipyridylium (dibromide)
   paraquat 1,1'-dimethyl-4,4'-bipyridylium (dichloride)
   difenzoquat 1,2-dimethyl-3,5-diphenylpyrazolium (methyl
      sulphate)

E. *Organophosphorus compounds*
   glyphosate N-(phosphonomethyl)glycine
   fosamine ammonium ethylcarbamoylphosphonate

F. *Synthetic growth-regulators* (hormones)
   i. Phenoxyacetics
      2,4-D 2,4-dichlorophenoxyacetic acid
      MCPA 4-chloro-2-methylphenoxyacetic acid

   ii. Phenoxypropionics
      dichlorprop 2-(2,4-dichlorophenoxy)propionic acid
      mecoprop (CMPP) 2-(4-chloro-2-methylphenoxy)propionic
      acid

   iii. Phenoxybutyrics
      2,4-DB 4-(2,4-dichlorophenoxy)butyric acid
      MCPB 4-(4-chloro-2-methylphenoxy)butyric acid

   iv. Benzoics
      dicamba 3,6-dichloro-2-methoxybenzoic acid
      2,3,6-TBA 2,3,6-trichlorobenzoic acid

v. Picolinics

    clopyralid (3,6-DCPA) 3,6-dichloropicolinic acid
    picloram 4-amino-3,5,6-trichloropicolinic acid
    triclopyr 3,5,6-trichloro-2-pyridyloxyacetic acid

G. *Nitriles*

    bromoxynil 3,5-dibromo-4-hydroxybenzonitrile
    chlorthiamid 2,6-dichlorothiobenzamide
    dichlobenil 2,6-dichlorobenzonitrile
    ioxynil 4-hydroxy-3,5-diiodobenzonitrile

H. *Carbamates and thiocarbamates*

    asulam methyl N-(4-aminobenzenesulphonyl)carbamate
    barban 4-chloro-2-butynyl-N-(3-chlorophenyl)carbamate
    EPTC S-ethyl N,N-di-*n*-propylthiocarbamate
    karbutilate m-(3,3-dimethylureido)phenyl *t*-butylcarbamate
    triallate S-2,3,3-trichloroallyl-N,N-diisopropylthiocarbamate

I. *Substituted ureas*

    chlorsulfuron N-(2-chlorobenzenesulphonyl)-N-(4-methoxy-6-
        methyl-1,3,5-triazin-1-yl) urea
    diuron N'-(3,4-dichlorophenyl)-*N,N*-dimethylurea
    ethidimuron N'-(5-ethylsulphonyl-1,3,4-thiadiazol-2-yl)-*N,N*-
        dimethylurea
    fluometuron N'-(3-trifluoromethylphenyl)*N,N*-dimethylurea
    linuron N'-(3,4-dichlorophenyl)-*N*-methoxy-*N*-methylurea
    metobromuron N'-(4-bromophenyl)-*N*-methoxy-*N*-methylurea
    metoxuron N'-(3-chlorophenyl)-*N*-methoxy-*N*-methylurea
    monuron N'-(4-chlorophenyl)-*N,N*-dimethylurea
    thiazafluron 1,3-dimethyl-(5-trifluoromethyl-1,3,4-thiadiazol-
        2-yl)urea

J. *Triazines*

    atrazine 2-chloro-4-ethylamino-6-isopropylamino-1, 3, 5-tria-
        zine
    ametryne-4-ethylamino-6-isopropylamino-2-methylthio-1,3,5-
        triazine
    cyanazine 2-chloro-4-(1-cyano-1-methylethylamino)-6-ethyla-
        mino-1,3,5-triazine
    dimethametryn 2-(1,2-dimethylpropylamino)-4-ethylamino-6-
        methylthio-1,3,5-triazine
    prometryne 4,6-bisisopropylamino-2-methylthio-1,3,5-triazine
    simazine 2-chloro-4,6-bisethylamino-1,3,5-triazine
    terbutryne 4-ethylamino-2-methylthio-6-*t*-butylamino-1,3,5-
        triazine

K. *Uracils and pyridazines*

   bromacil 5-bromo-3-*sec*-butyl-6-methyluracil

   lenacil 3-cyclohexyl-5,6-trimethyleneuracil

   terbacil 5-chloro-6-methyl-3-*t*-butyluracil

   norflurazon 4-chloro-5-methylamino-2-(3-trifluoromethyl-phenyl) pyridazin-3-one

L. *Triazinones*

   hexazinone 3-cyclohexyl-6-dimethylamino-1-methyl-1,3,5-triazine-2,4-dione

   metribuzin 4-amino-6-*t*-butyl-3-(methylthio)-1,2,4-triazin-5(4*H*)-one

M. *Amides*

   alachlor 1-chloro-2,6-diethyl-*N*-(methoxymethyl)acetanilide

   butachlor *N*-(butoxymethyl)-1-chloro-2,6-diethylacetanilide

   methachlor 1-chloro-6-ethyl-*N*-(2-methoxy-1-methylethyl)-O-acetotoluidide

   proachlor 1-chloro-*N*-isopropylacetanilide

   propanil *N*-(3,4-dichlorophenyl) propionamide

N. *Dinitroanilines*

   dinitramine *N'*,*N'*-diethyl-2,6-dinitro-4-trifluoromethyl-*m*-phenylenediamine

   oryzalin 3,5-dinitro-*N4* *N4*-dipropylsulphanilamide

   pendimethalin *N*-(1-ethylpropyl)3,4-dimethyl-2,6-dinitro-benzeneamine

   trifluralin 2,6-dinitro-*N,N*-dipropyl-4-trifluoromethylaniline

O. *Diphenyl ethers*

   fluorodifen 2,4-dinitro-4-trifluoromethyldiphenylether

   oxyfluorfen 2-chloro-4-trifluoromethylphenyl-3-ethoxy-4-nitro-phenylether

P. *Phenoxy propionates and phenylaminopropionates*

   diclofop-methyl methyl 2-[4-(2,4-dichlorophenoxy) phenoxy] propionate

   L-flamprop-isopropyl L-isopropyl *N*-benzoyl- *N*-(3-Chloro-4-fluorophenyl)-2-aminopropionate

   fluazifop-butyl butyl 2-[4-(5-trifluoromethyl-2-pyridyloxy) phenoxy] propionate

Q. *Hydroxylamine derivatives*

   alloxydim 2-[1-(*N*-allyloxyamino)butylidene]-4-methoxycar-bonyl-5,5-dimethylcyclohexane-1,3-dione

(xiii)

sethoxydim 2- [1-(ethoxyamino)butylidene]-5-(ethylthiopro-
pyl)-cyclohexane-1,3-dione

R. Miscellaneous herbicides
amitrole (ATA) 3-amino-1,2,4-triazole
bentazon 3-isopropyl-2,1,3-benzothiadiazin-4-one-2,2-dioxide
chlorthal-dimethyl 2,3,5,6-tetachloro terephthalic acid dime-
thyl ester
oxadiazon 3-(2,4-dichloro-5-isopropoxyphenyl)-5-t-butyl-1,3,4-
oxadiazolin-2-one

S. Safeners (antidotes)
naphthalic anhydride 1,8-naphthalic anhydride
R-25788 N, N-diallyl-2,2-dichloroacetamide

T. Soil sterilants
metham-sodium sodium N-methyldithiocarbamate
methyl bromide
methyl isothiocyanate (in mixture with dichloropropene and
dichlopropane).

*Note:
List of chemicals produced from information provided by the chemical companies. The list has been made as complete as possible but omission does not
necessarily indicate that a chemical is unavailable nor inclusion that it is currently being marketed.

Abbreviations

cm	centimetre
diam	diameter
ha	hectare
kg	kilogram
l	litre
m	metre
mm	millimetre
sq	square

All herbicide application rates are given in terms of active ingredient
(the same as acid-equivalent for acid herbicides such as the growth-
regulators).

WATER WEEDS

APONOGETONACEAE

Aponogeton abyssinicus Hochst, ex A. Rich.

DESCRIPTION
A soft-stemmed, fast-growing perennial growing in shallow water with leaves floating on the surface and paired spikes of mauve flowers on stems projecting above the surface. The plant grows from a tuber buried in the mud. The leaves are borne on stalks up to 50 cm long and are very variable in shape and size, up to 16 cm long × 5 cm wide, more or less parallel sided with a blunt tip and with a wide central vein and four or more smaller, parallel veins. The flowering stems project a few centimetres above the water and at the tip bear two diverging spikes up to 6 cm long × 1 cm wide. The flowers are numerous, more or less densely crowded and directed outwards on all sides of the axis. Each flower consists of two petals, six stamens and three ovaries.

Distribution and importance
A species restricted to East Africa, *A. abyssinicus* is widespread in wet situations in Kenya, Uganda, northern and western Tanzania from sea level to 2,700 m. It is locally common in shallow water at the edges of lakes and rivers, in irrigation channels and temporary ponds. It is reported to be a weed of importance in rice on the Mwea Irrigation Scheme.

Methods of control
No information is available on the control of this species. With the related *A. distachyus* in New Zealand it is reported that pelleted 2,4-D or dichlobenil gave successful control but that the tubers resprouted after application of diquat. The tubers of *A. abyssinicus* would also be expected to resprout following application of a contact herbicide.

ARACEAE

Pistia stratiotes L.   Nile Cabbage, Water Lettuce

DESCRIPTION
A free-floating plant sending down long, feathery roots into the water and usually bearing several runners, each with a small plant at the end.

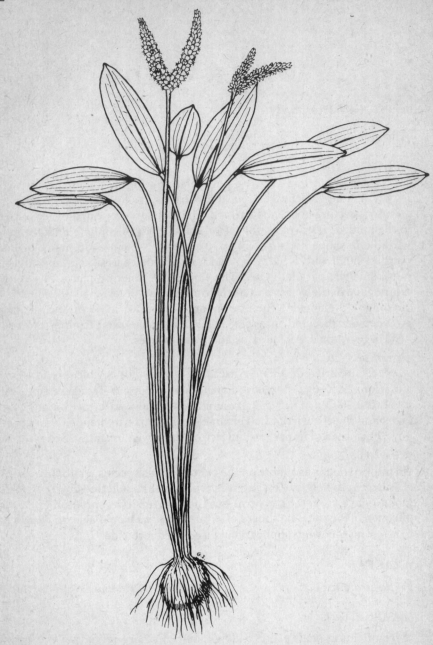

Figure 1. Aponogeton abyssinicus

It consists of a rosette of pale green, wedge-shaped leaves, up to 12 cm long × 5 cm wide. The prominent veins are arranged in a fan-like formation and both leaf surfaces are covered with fine hairs. The minute flowers are very inconspicuous and are surrounded by green, tubular sheaths, arising in the leaf axils. The plant proliferates rapidly by means of its runners.

Distribution and importance

An indigenous plant common in still or slow-moving water throughout East Africa and the tropics generally. It frequently forms dense masses among reeds along the banks of rivers and, on occasion, has been mistaken for *Eichhornia*, the Water Hyacinth. *Pistia*, however, is a smaller plant which does not multiply so rapidly as *Eichhornia* and rarely occurs in quantities large enough to be troublesome. It has been reported as causing trouble on the Pangani river in Tanzania by blocking up the inlet grilles of the Pangani Falls hydroelectric station.

Methods of control

Pistia has only a limited susceptibility to herbicides of the 2,4-D type, partly because the hairy leaf surface is difficult to wet with aqueous sprays. Recent work in the U.S.A. however has shown that good control can be obtained with diquat at 1 kg per ha. and success with this chemical has also been obtained in Nigeria. It is recommended that stock should be kept away from treated water for 7 days after application. In the Philippines good control in rice paddies is also reported with glyphosate applied before transplanting.

LEMNACEAE
Wolffia arrhiza (L.) Wimm.   Duckweed

DESCRIPTION

A minute, rootless, floating plant consisting of a thickened, more or less ovoid thallus up to 1.5 mm long, which produces daughter plants by budding. The flowers are small and very rarely seen. When they do occur, they are produced singly in a hollow on the upper side of the thallus and are reduced to a single stamen and an ovary containing a single ovule.

Distribution and importance

Widely distributed and not uncommon in ponds, dams and still water generally. Of minor importance as a weed, but suspected of causing tainting when growing in reservoirs supplying coffee factories in parts of Kenya.

4

Figure 2. Pistia stratiotes

Methods of control
The use of chemicals has not been attempted in East Africa, but trials in New Zealand suggest that control can be achieved readily with diquat,

and in India simazine at a dose of 5 kg per ha is reported to be effective. Before any chemical can be contemplated for application to water used for coffee processing, however, extensive testing will be needed to ensure the absence of toxicity or taint in the treated water. Biological control by fish or ducks might well be more practicable in such situations.

Figure 3. Wolffia arrhiza

NAJADACEAE

Najas graminea Delile

DESCRIPTION

A yellowish-green submerged plant rooted in the mud in shallow water, with tufts of very fine leaves at the nodes. The slender, brittle stems are up to 50 cm long, with leaves mostly in threes lower down the stem and tufted in the upper part where the side branches are short. The leaves are up to about 25 mm long and thread-like, with minute teeth on the margins and two small projections at the base which clasp the stem. The flowers are minute, green and stalkless in the leaf axils. Male and female flowers are separate, the male consisting of a single stamen, the female of a single carpel containing one ovule.

Distribution and importance

A species of widespread occurrence in the tropics and sub-tropics, commonly growing in lakes and irrigation ditches, sometimes as a weed in irrigated rice. In Kenya reported as a weed of significance in the Mwea Irrigation Scheme, where studies have shown up to 60% of the weed seeds in the soil to consist of this species. Although it may be present in large numbers, *N. graminea* does not appear to be a strongly competitive species.

Methods of control

Few attemps have been made to control this weed and no specific information is available on the effects of chemicals. As it grows under water,

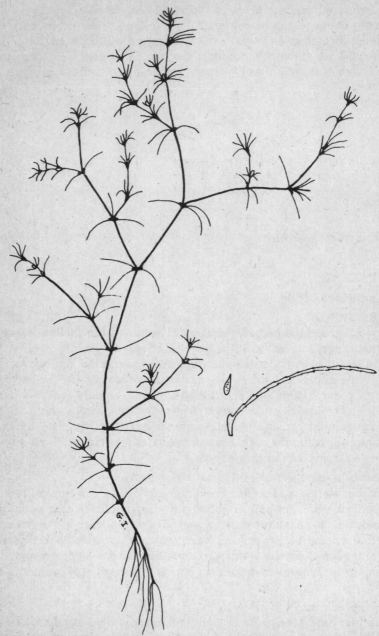

Figure 4. Najas graminea

contact herbicides such as propanil, the most commonly used rice herbicide, would not be expected to give control. Formulations containing a residual chemical, such as butachlor or oxadiazon, offer a better chance of being effective in formulations which can be applied to the flooded crop.

PONTEDERIACEAE
Eichhornia crassipes (Mart.) Solms   Water Hyacinth

DESCRIPTION
A floating plant with a thick rhizome and long, purple, feathery roots which hang free in deep water or anchor the plant to the mud in the shallows. Daughter plants are produced at the ends of runners. The smooth fleshy leaves are bright green and arranged in rosettes with a swollen stalk 5—30 cm long (up to 1 m long when growing densely under hot, humid conditions) and a broad, pointed blade. The pale lilac flowers are up to 5 cm across and form a spike-like inflorescence. There are 6 petals, the upper one marked with blue and yellow, 6 stamens and a 3-segmented ovary. After flowering, the inflorescence bends down into the water and the seed sinks to the bottom. It is not known whether fertile seed is set in East Africa.

Distribution and importance
Originally introduced for its decorative value, *Eichhornia* is now scheduled as a noxious weed throughout Eastern Africa and its culture is illegal. A native of tropical America, it is now established in nearly all tropical countries and where rivers are important for transport its remarkable powers of spreading make it a very serious pest. The weed is particularly serious on the Congo river where thousands of miles are infested and where enormous expenditure has failed to give satisfactory control, but it is also now well established on the Nile and causes considerable interference with river transport in the Sudan.

In East Africa, although probably present in ponds and dams for some time, the first place that *Eichhornia* was observed to have got out of hand was on the Sigi river, near Tanga in Tanzania, in 1955. In 1959 it also appeared on the nearby Pangani river in the neighbourhood of Korogwe and spread rapidly downstream. At present these are the only two major infestations known but the weed has already been reported as far up the Nile as Nimule, on the Sudan border, and if it ever becomes established in the swamps and waterways of Uganda it is likely to become much more of a problem in the next few years than it is at present.

It is important that the spread of *Eichhornia* should be closely watched and that any fresh infestations discovered should be dealt with before the weed has spread too widely for eradication to be possible. In the Congo two plants have been observed to produce 1,200 daughter plants in four months and, in a really dense infestation, the mass of vegetation can be thick enough to support the weight of a man.

Methods of control

Eichhornia is susceptible to 2,4-D and other herbicides of the growth-regulator type and rates of the order of 1 kg per ha will give a considerable degree of control. Because of the importance of obtaining 100 per cent kill, however, and because it is often difficult to obtain even coverage of water weed infestations, higher rates of 3-6 kg per ha are generally recommended. Amine formulations of 2,4-D are probably the best materials to use and should be applied in sufficient water to obtain complete coverage while avoiding run-off. Chemicals of the 2,4-D type have a low level of toxicity to animals and the suggested rates are harmless to fish.

When the weed is growing very thickly some plants are usually sheltered by others so that they escape the spray and a second application is needed. In deep water, however, the whole mass of dying vegetation often sinks to the bottom, carrying any surviving plants with it and killing them. It is important that a close watch should be kept on the area after treatment in order to guard against re-invasion and this is particularly important where fertile seed is produced as the seeds can remain dormant for several years. Glyphosate has also given good control of *Eichhornia* in the USA. At the present time, however, 2,4-D is still to be preferred on a cost basis.

Although *Eichhornia* is susceptible to herbicides, attempts to eradicate the weed have rarely been entirely successful because of the difficulty of reaching all the areas of marsh and swampy thicket in which the plant becomes established. Especially bad as centres of re-invasion are the beds of aquatic grasses often fringing the banks of rivers and particular attention has to be paid to such places in any control campaign.

Attention is currently being directed towards finding a suitable biological method of controlling *Eichhornia*. Various organisms appear to offer possibilities, such as manatees in Guyana, and geese and weevils (*Neochetina eichhorniae*) in the USA. Promising fungal control agents have also been found, but large-scale biological control has yet to be achieved in Africa.

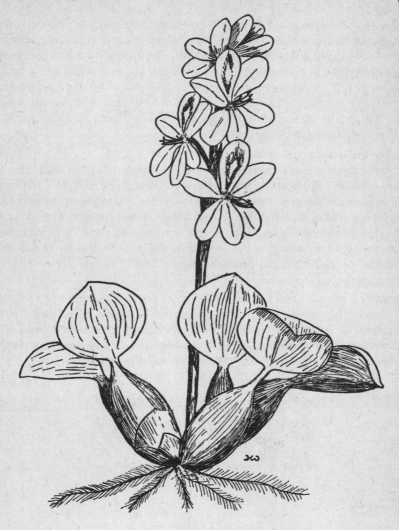

Figure 5. Eichhornia crassipes

SALVINIACEAE

Azolla africana Desv.   Water Velvet

DESCRIPTION

A free-floating, aquatic fern. Plants small, more or less triangular in shape, up to about 5 cm long, bluish green or sometimes brownish. The main stem bears up to 5 pairs of branches, arranged alternately, from which grow tufts of fine roots and numerous alternate, sessile leaves, about 1 mm long and densely overlapping towards the ends of the branches. Each leaf is divided into two lobes, the upper floating and green, the lower thin, submerged, colourless and bearing the spore producing structures. The aerial parts of the leaves are covered with minute hairs which make the surface very difficult to wet. The plant multiplies rapidly and is often found in dense masses.

Distribution and importance

An indigenous species distributed widely through Africa, *Azolla* is commonly found in dams and other areas of still water. In East Africa it is of minor importance as a weed, but is sometimes a nuisance in reservoirs supplying coffee factories. In West Africa it has been reported as a troublesome weed of rice fields along the Gambia river. It is included here mainly to illustrate the distinction from the more important *Salvinia molesta*, with which it may possibly be confused.

Methods of control

Chemicals are not known to have been tested in East Africa, but on similar species in New Zealand both diquat and paraquat have been found effective. In South Africa a spray of diesel oil has also given control. As mentioned in the section on *Wolffia*, however, chemical control in water to be used for coffee processing is unlikely to be practicable.

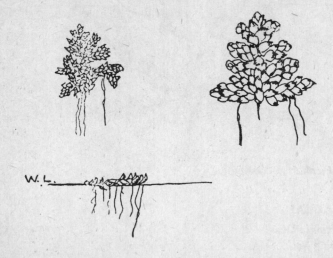

Figure 6. Azolla africana

SALVINIACEAE

Salvinia molesta Mitch.   Salvinia

DESCRIPTION

A free-floating, branched, aquatic fern capable of very rapid growth. Individual plants are up to 30 cm long with numerous leaves forming a mat up to about 2 cm thick, but under favourable conditions dense masses of plants develop, which can cover large areas with a much thicker layer. The leaves are borne in groups of three, two green, broadly oval, undivided, aerial leaves up to 2.5 cm long and a brown, much divided, submerged leaf up to 15 cm long, with the appearance and function of roots but also bearing the spore forming structures. The leaves of young plants growing in open water float flat on the surface; later as the plants become more crowded they fold along the midrib so that the two halves become vertical with their upper surfaces facing each other. The upper surfaces are densely covered with long, club-shaped hairs and are very difficult to wet.

Distribution and importance

A native of South America, *Salvinia* is often grown in aquaria and has been introduced to a number of tropical countries. It is a particularly serious pest in Sri Lanka, where it causes much choking of waterways and flooding. In Africa it achieved prominence in 1959 when it was discovered to be firmly established in the rising waters of Lake Kariba, and within a few years was covering hundreds of square miles with a mat up to 25 cm thick.

It was first recorded as a weed in East Africa in 1957 when it was accidentally introduced into a dam near Kitale, in Kenya. Within a few months the *Salvinia* had multiplied to such an extent that the surface of the water was completely covered and a boat could only be moved through the water with the aid of a rope pulled from the bank. As the overflow from the dam eventually led into Lake Victoria there was a danger of the weed spreading further, but combined aerial spraying, hand removal and drainage appear to have cleared up the infestation.

Salvinia was subsequently scheduled as a noxious weed in the East African countries and its cultivation forbidden by law, but this did not prevent its appearance in Lake Naivasha in 1964, where it still persists despite attempts at eradication.

Methods of control

In irrigation channels, and other waterways where the water level can be controlled, desiccation of the weed through drainage is an obvious method of control, and one which is extensively used in Sri Lanka. Where

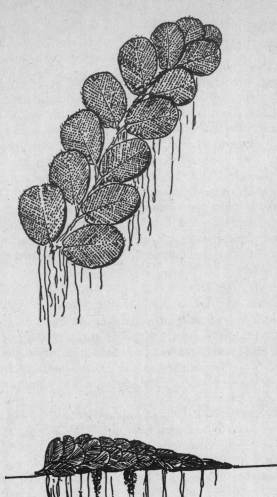

Figure 7. Salvinia molesta

drainage is not possible chemicals may be necessary. The first herbicide found to give control was pentachlorophenol (PCP), applied as an emulsified oil solution by spraying. More recently, experiments with a wide range of herbicides at Lake Kariba have shown *Salvinia* to be sensitive to paraquat, which at a dose of 1 kg per ha is highly effective and less harmful to fish. In Queensland an oil-based spray containing diuron has also been effective.

Experience in Kenya, as elsewhere, is that where drainage is impossible, eradication cannot be achieved because spray applications always miss pockets of weed concealed among emergent aquatic vegetation. *Salvinia* control operations, therefore, must make allowance for regular inspection and retreatment of infested areas.

A number of insects which can feed on *Salvinia* are being tested as biological control agents. Among these *Paulinia acuminata*, an aquatic grasshopper, appears to have given effective control on Lake Kariba. This insect has also been tested at Naivasha.

Figure 8. Cyperus rotundus

GRASSES, SEDGES ETC.

CYPERACEAE

Cyperus rotundus L.   Nutgrass, Watergrass

C. tuberosus Rottb.
   (= *C. rotundus* L. subsp. *tuberosus* (Rottb.) Kük.)

C. esculentus L.

C. grandibulbosus C.B.Cl.

C. blysmoides C.B.Cl.
   (= *C. bulbosus* Vahl var. *spicatus* Boeck.)

C. maranguensis K. Schum.

C. rigidifolius Steud

DESCRIPTION

Cyperus rotundus is a leafy perennial usually 15 —30 cm high smooth, 3-angled stem and narrow grass-like leaves growing from a swollen shoot base. Slender runners are produced below ground and form chains of black, irregular shaped tubers up to 2 cm long. These tubers often sprout to produce new shoots while still attached to the original plant. The inflorescence, subtended by several leaf-like bracts, grows from the top of the stem and consists of a number of slender, spreading branches of unequal length, near the ends of which are clustered the narrow, flattened, dark reddish-brown spikelets 1 —2 cm long. Each spikelet is made up of 10 —30 closely crowded florets which ripen into black 3-angled nutlets.

   C. tuberosus, sometimes regarded as a subspecies of *C. rotundus*, has a similar growth habit but is a larger plant, up to 60 cm high, with spikelets which are usually broader and of a lighter brown colour.

   C. esculentus. Above ground this species is also similar to *C. rotundus* in general appearance but differs in having yellowish-brown spikelets and in the finely tapered leaf tips. Below ground the rounded tubers at the ends of runners are an additional distinguishing feature.

   C. grandibulbosus. A bulb-forming species with yellowish-brown spikelets, differing from *C. esculentus* in the shorter, more crowded inflorescence branches and in the way the dark-brown bulbs, develop at intervals along the runners. There are two varieties of this species, one with a more compact inflorescence than the other (see Figure 8).

C. blysmoides usually only grows about 15 cm high, the narrow leaves arising from a black bulb and often longer than the flowering stem. The inflorescence consists of 3 —6 narrow, reddish-brown spikelets in an un-branched spike. A small, pointed bulb develops at the end of each slender, underground runner.

C. maranguensis. The above-ground parts are similar to *C. rotundus,* except for the very narrow spikelets of a dark grey-green colour. Also differs in being a tufted plant without tubers.

C. rigidifolius. The shoots arise at frequent intervals from a thick, woody rhizome just below ground level. Leaves and stems are tough and difficult to cut, the inflorescence of blackish-brown spikelets crowded into dense clusters.

In addition to the species described here, various other *Cyperus* species and related plants occur as weeds in East Africa. Terry (1978) recorded 57 species in total, 10 of which cause serious problems and a further 9 minor problems. For identification of other species the papers by Terry (1976) or Napper (1966) should be consulted.

Distribution and importance
C. rotundus has the reputation of being the 'world's worst weed' and is the commonest and most troublesome *Cyperus* in most parts of East Africa, except for southern and western Tanzania. It occurs in all types of crop up to about 2000 m and is particularly serious in coffee, sugar and maize. *C. tuberosus* causes similar problems but is more restricted in its distribution, occurring mainly in low-lying and coastal areas.

C. esculentus appears to grow best at higher altitudes and is an important weed in many crops of the Kenya highlands, northern Tanzania and the Southern highlands. In maize it is reported to be encouraged by techniques of minimum tillage. *C. grandibulbosus* can also be a serious problem but is largely restricted to irrigated crops in the Mwea Tebere and Hola districts of Kenya.

C. blysmoides is the watergrass which infests wheat in the West Kilimanjaro area and is also common in coffee and other crops in the Kenya highlands. *C. maranguensis* is another highland species common in parts of Kenya and northern Tanzania but not generally considered a very serious problem. *C. rigidifolius* is more common as a weed of pastures and uncultivated land in highland areas than in crops but has been reported as troublesome in young tea at Mufindi.

The perennial *Cyperus* species are very persistent weeds because, although the tops may be killed by cultivation or spraying, the bulbs or tubers are little affected and soon produce new shoots. Very dense infes-

tations can build up in a few years. With *C. rotundus* up to 1000 shoots/
m have been noted, while with *C. blysmoides* densities may rise to
2000/m , each plant capable of producing 40 or more new bulbs.

Methods of control

With *C. rotundus, C. esculentus* and, probably, some other species
2,4-D or MCPA at relatively high rates give some degree of control. The
tops of young plants are killed and regrowth may be delayed but the
tubers themselves are little affected. In newly planted sugar cane this
degree of control may delay weed growth long enough for the crop to
establish and shade out the later emerging shoots. In most other crops
the effects of such treatment are not sufficiently long lasting.

Among other leaf-absorbed chemicals, glyphosate at higher rates
(preferably as a split application of 2 kg/ha followed by another 1 kg/ha
on any regrowth that appears) gives effective control and can be used as
a pre-planting or directed treatment. Paraquat also kills the tops but
must be applied repeatedly at about 4-week intervals to give lasting control.

Among residual chemicals, high rates of various urea, uracil and triazine herbicides can be used to control *Cyperus* species in non-agricul-
tural land. In a few crops they can also be used selectively. Bromacil, for
example, may be applied in citrus and pineapple; ametryne is recommended for *Cyperus* control in sugar cane, coffee and tea.

The most reliable control in annual crops, including maize and dwarf
French beans, has been obtained with EPTC. This chemical is volatile
and must be incorporated into the soil soon after application to be effective. It must, therefore, be used as a pre-planting treatment and in
maize, the formulation which includes a safener must be used. More
sensitive crops should not be sown for at least 2 months after applying
EPTC. An alternative chemical, which can be used to control *C. esculentus* in maize and various other crops on certain soil types is metoxuron. This material is partially leaf absorbed but is most effective
against *C. esculentus* when applied before either weed or crop emergence and when sufficient rain follows application to carry the chemical
into the soil. EPTC is reported to be less effective against *C. blysmoides*
but there is little published information on perennial *Cyperus* species
other than *C. rotundus* and *C. esculentus*.

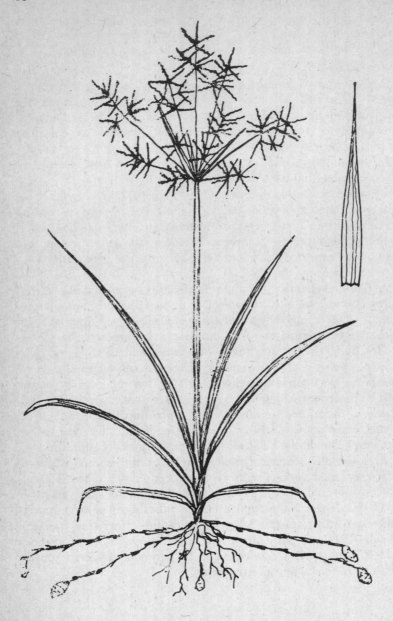

Figure 9. Cyperus esculentus

Figure 10. Cyperus grandibulbosus

CYPERACEAE

Cyperus difformis L.

DESCRIPTION

This weed differs from the other *Cyperus* species described in being an annual, without tubers, bulbs or rhizomes. It is a tufted plant up to 50 cm high and is readily distinguished by the branches of the inflorescence terminating in dense, spherical clusters of spikelets 6—12 mm in diameter and green or reddish-green in colour.

Distribution and importance

C. difformis is a serious weed of paddy rice throughout the tropics. It occurs widely in wet places in East Africa and, although not often recorded as an important problem, has the potential of becoming serious as a rice weed. Infestations can build up rapidly because the plant produces large quantities of seed, which can germinate at any time of year, and it completes its life cycle in 6—8 weeks, so that several generations can grow in one year.

Methods of control

Because of its lack of underground perennating organs *C. difformis* is easier to control than such species as *C. rotundus* and *C. esculentus*. As a post-weed-emergence treatment bentazone may be used in transplanted rice and has given good results in Tanzania in combination with propanil. Pre-weed-emergence chemicals are also available for use in rice including butachlor and oxadiazon, both of which are available as mixtures with propanil.

CYPERACEAE

Kyllinga bulbosa P. Beauv.   Creeping Sedge, Watergrass

K. erecta Schum.

K. odorata Vahl var. *major* (C.B.Cl.) Chiov.

DESCRIPTION

Kyllinga bulbosa is a perennial with narrow grassy leaves and stems up to 30 cm high arising from a swollen shoot base, from which grow runners giving rise to further shoots at intervals. The leaves have a distinctive aromatic smell when crushed. At the top of the flowering stem are several downwardly directed, leafy bracts and a densely clustered, white flower head made up of numerous spikelets arranged in several sessile, compacted spikes. Individual spikelets are cylindrical as opposed to *Cyperus* species where they are flattened.

x ⅖

x ⅘

Figure 11. Cyperus difformis

$x\frac{4}{5}$

$x\frac{2}{5}$

Figure 12. Kyllinga bulbosa

K. erecta has a similar habit of growth but differs in its tough, reddish-brown, creeping rhizomes, from which numerous shoots arise at close intervals, and in its spherical, yellowish flower heads 4—8 mm across subtended by three leafy bracts.

K. odorata var. *major* is a tufted plant with very short rhizomes. It can grow to 50 cm high and has a white flower head consisting of a sessile, central spike about 12 mm long × 9 mm wide with three smaller, lateral spikes at its base.

Distribution and importance

These three *Kyllinga* species are widespread and common throughout the region at altitudes above 800 m. *C. bulbosa* is a common weed of lawns and is also important in young tea at Kericho. *K. erecta* is especially prevalent in the north of Tanzania where it often forms a dense mat which excludes the more desirable lawn grasses. By itself it forms a turf of sorts but is unsightly during the dry season when it turns brown. *K. odorata* var. *major* occurs as a minor weed of various highland crops and is often a constituent of Kikuyu grass lawns.

Methods of control

Well established *Kyllinga* is little affected by selective herbicides and the only satisfactory method of dealing with a heavy infestation in a lawn is to cultivate, remove the rhizomes mechanically and replant. Dense masses of seedlings often appear in replanted lawns and good control of these can be achieved with MCPA or 2,4-D. On paths and other areas where bare ground is required such non-selective chemicals as glyphosate would be expected to be effective.

GRAMINEAE

Annual Grasses

Numerous species of annual grass occur as weeds in crops, gardens and uncultivated land. Some of the more important are described here but space only permits the inclusion of a limited selection of those likely to be encountered. Comprehensive keys for the identification of East African grasses are available and should be consulted. The following are particularly useful: Bogdan 1958, Harker and Napper 1960, Napper 1966, and the 3 Gramineae volumes (1970, 1974 and 1982) in the Flora of Tropical East Africa.

The following list includes the commoner annual grasses occurring as weeds and indicates those which are described and illustrated later.

Aristida adscensionis L.		Widespread
Avena fatua L.	Wild Oat	Highlands (Fig. 15)
A. sterilis L.	Wild Oat	Highlands
Bromus diandrus Roth	Brome Grass	Highlands
B. pectinatus Thunb.		Highlands
Chloris pycnothrix Trin.	False Star Grass	Widespread
Dactyloctenium aegyptium (L.) Willd.	Crows-foot Grass	Widespread (Fig. 13)
Digitaria ternata (A. Rich.) Stapf		Widespread Highlands
D. velutina (Forsk.) Beauv.		Widespread
Echinochloa colona (L.) Link	Barnyard Grass	Widespread (Fig. 20)
Eleusine indica (L.) Gaertn.	Wild Finger-millet	Widespread (Fig. 21)
Eleusine multiflora A. Rich.		
Eragrostis tenuifolia (A. Rich) Steud.		Widespread
Harpachne schimperi A. Rich.		Widespread (Fig. 14)
Lolium temulentum L.	Darnel	Highlands (Fig. 24)
Oryza barthii A. Chev.	Wild Rice	Tanzania
O. punctata Steud.	Wild Rice	Lowland (Fig. 25)
Paspalum scrobiculatum L.		Widespread (Fig. 27)
Rhynchelytrum repens (Willd.) C.E. Hubbard	Red-top Grass	Widespread
Rottboellia cochinchinensis (Lour.) Clayton (= *R. exaltata* L.f.)	Guinea-fowl Grass	Widespread (Fig. 29)
Setaria pumila (Poir.) Roehm. & Schult. (= *S. pallide-fusca* (Schum.) Stapf & Hubbard)		Widespread
S. verticillata (L.) Beauv.	Love Grass	Widespread (Fig. 30)
Snowdenia polystachya (Fres.) Pilg.	Abyssinian Grass	Highlands (Fig. 31)
Sorghum arundinaceum (Desv.) Stapf (= *S. verticilliflorum* (Steud.) Stapf		Widespread
Urochloa panicoides Beauv.		Widespread

Figure 13. Dactyloctenium aegyptium

Methods of control

Control of annual grass weeds by cultivation or spraying usually presents no problems in perennial crops but is more difficult in arable crops, especially in annual grass crops.

Many soil-applied chemicals control grasses effectively in the early stages of growth and selective pre-emergence treatments for a wide range of crops are based on herbicides of the thiocarbamate (e.g. EPTC), amide (e.g. alachlor), dinitroaniline (e.g. trifluralin) and substituted urea (e.g. diuron) groups. The triazine herbicides (e.g. atrazine) also have pre- and early post-emergence uses in maize and a number of perennial crops and give effective control of many annual grasses. Species belonging to the sub-family *Panicoideae* (e.g. *Digitaria, Echinochloa, Rhynchelytrum, Rottboellia, Setaria, Sorghum* and *Urochloa* species), however, are resistant to triazines so that, where grasses of this type predominate, mixtures with other types of herbicide must be applied.

The foliage acting herbicide paraquat is a very useful material for controlling seedling or older annual grass weeds. It is non-selective, so that it also controls most annual broad-leaved weeds and must be kept off the crop foliage. Where it is possible to direct the spray so as to cover the weeds and avoid the crop, as in bush and tree crops, paraquat can be used very successfully and, being inactivated on contact with the soil, has no residual activity. Glyphosate is another effective grass killing herbicide which is non-selective and non-residual and can be used in many of the same situations as paraquat. This product also gives good control of many of the more difficult perennial grass weeds.

GRAMINEAE

Avena fatua L. (s.l.)   Wild Oats

A. sterilis L.   Wild Oats

DESCRIPTION

Stout, broad-leaved annual grasses 1 m or more high, very much like cultivated oats in general appearance, with branched, spreading inflorescences. The pendulous spikelets are 1.8 —2.5 cm long in *A. fatua* and 3.8 —5 cm in *A. sterilis*. They are made up of 2 —5 florets enclosed by a pair of papery bracts (glumes), and each floret bears a long, bent bristle. The seeds of *A. fatua* are 1.2 —1.8 cm long and of *A. sterilis* 2.5 —3.8 cm. Both species are readily distinguished from cultivated oats by the presence of long, silky hairs round the base of the seed.

Figure 14. Harpachne schimperi

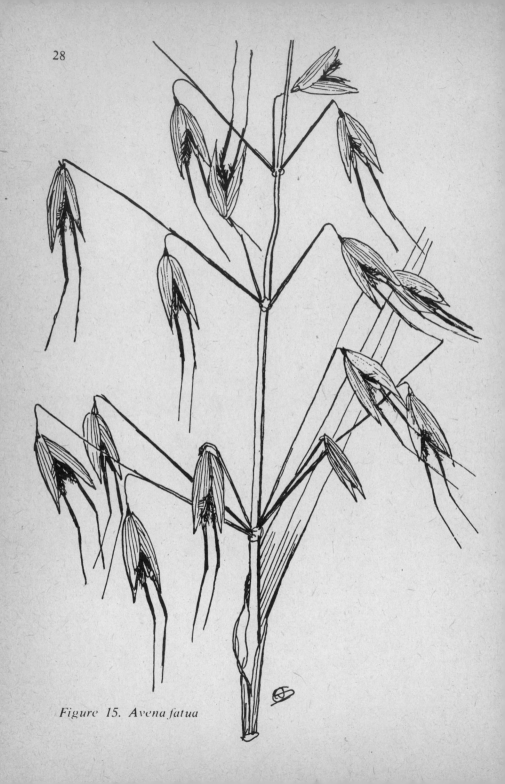

Figure 15. Avena fatua

Distribution and importance

These troublesome weeds have been introduced from Europe and are on the increase in the more temperate parts of East Africa. They occur mainly as cereal weeds in the Kenya highlands and are particularly abundant in the Ol Kalou and Ol Joro Orok areas. *A. fatua* has also been recorded from the Southern Highlands of Tanzania.

Methods of control

The seed of wild oats can stay dormant in the soil for a number of years so that, once introduced, eradication is virtually impossible. In un-infested areas where temperate cereals are grown, therefore, all possible precautions should be taken against introducing the weed in unclean seed. Where it appears for the first time in small numbers it may be pos-sible to prevent spread by pulling out the plants by hand and destroying them before the seed has shed. Once the weed is established, the in-festation can be reduced by cultivations timed to destroy the young plants before they produce seed. If sowing of a cereal is delayed for some time after preparing the seedbed, the wild oats are encouraged to germinate and can then be destroyed by another cultivation before dril-ling. A similar procedure can be adopted after harvest before sowing the next crop and there are obvious opportunities for cultivation in row crops.

Several chemical treatments have also been developed. Triallate, for example, can be used in wheat barley and peas when applied before sowing and shallowly incorporated into the soil. Various post-emergence treatments are also available for use in wheat or barley which require careful timing. Barban is applied when the weeds are in the 1—2½-leaf stage, difenzoquat at the 3—5-leaf stage. Flamprop-isopropyl is recom-mended from the mid-tillering to the 1—2-node stage of the crop but the weeds must not have passed the first-node stage. Metoxuron can be used from the end of the 2-leaf stage of the crop until the start of tillering.

GRAMINEAE

Cymbopogon caesius (Hook. & Arn.) Stapf   Lemon Grass
   (= *C. excavatus* (Hochst.) Burtt-Davy)

C. nardus (L.) Rendle
   (= *C. validus* Burtt-Davy, *C. afronardus* Stapf)

DESCRIPTION

Coarse, tufted perennial grasses 1—2 m high, the leaves having a distinc-tive aromatic scent when crushed. The inflorescences are much branch-ed, the branches diverging little from the main stem, so that a relatively dense head is formed. The larger branches arise in the axils of reduced

Figure 16. Cymbopogon caesius

leaves, the smaller in the axils of sheathing, leaf-like bracts. Each of the smallest branches terminates in two elongated groups of spikelets about 1 cm long, the spikelets also being paired, one of each pair sessile, the other on a short stalk. The sessile spikelet is fertile, with a long, abruptly bent bristle, the stalked spikelet sterile and without a bristle.

The two species are similar in general appearance. *C. caesius* grows to a height of 0.6 – 1.2 m and has an erect inflorescence 8 – 15 cm long. The bristles on the spikelets are about 1 cm long and, in contrast with the other species, the outer scale of the sessile spikelet has a fine groove down the back.

C. nardus is usually larger and reaches 1 – 3 m in height. Its inflorescence is somewhat nodding and varies in length from 15 – 60 cm. The bristles are about 5 mm long and the outer scale of the sessile spikelet has no median groove.

Distribution and importance

Cymbopogon species are common constituents of the more open types of grassland over large areas of East Africa. Their importance lies in the fact that they are of very low grazing value and unpalatable to stock. *C. caesius* is the most widely distributed, being found at medium altitudes in Kenya, Tanzania and Uganda. *C. nardus* occurs mostly in areas of higher rainfall between 1100 – 2800 m often in areas where *Combretum* species are the dominant trees, and in upland grassland in the three countries.

Methods of control

Their densely tufted habit makes these grasses very resistant to fire and too frequent burning is one of the main causes of their increase. They are also encouraged by overgrazing. Strict control of burning and of the stocking rate is undoubtedly effective in limiting the spread of *Cymbopogon* species, but a system of management which will eradicate existing stands has yet to be developed. It is probable that the use of herbicides effective against grasses, such as dalapon or glyphosate, would facilitate the elimination of *Cymbopogon* from pastures. Hoeing out *C. nardus* in Uganda has resulted in a 30 per cent increase in growth rate of bullocks.

GRAMINEAE

Cynodon dactylon (L.) Pers.   Star Grass, Couch, Bermuda Grass

C. nlemfuensis Vanderyst

DESCRIPTION

Cynodon dactylon is a perennial grass with long slender runners which root at the nodes, and extensive underground rhizomes. The flowering stems are 8 —45 cm high, the leaves up to 15 cm long, often folded and stiff under dry conditions. By contrast with *Digitaria abyssinica* (from which *Cynodon* is not always readily distinguished when no flowers are present) there is no obvious membranous ligule where the leaf blade joins the sheath. The inflorescence consists of several (usually 4 —5) slender, purplish spikes up to 10 cm long, arising in a star-like arrangement from the end of the stem. Along the underside of the spikes the numerous, sessile spikelets are arranged in two rows. Each spikelet contains a single floret 2.5 mm long, and there are no bristles (a distinguishing feature from *Chloris pycnothrix*, an annual with an otherwise very similar inflorescence). *C. dactylon* is an extremely variable species, some varieties being employed as lawn grasses, others being more suitable for grazing.

C. nlemfuensis resembles a coarse variety of *C. dactylon*, with stouter, tough runners, stems up to 60 cm high and broader leaves. Its main distinguishing feature is that it does not develop underground rhizomes. The runners of this species grow extremely rapidly and the rooting nodes may be 1 m or more apart.

Distribution and importance
C. dactylon is a weed of world-wide occurrence. It is not clear at the moment whether it or *C. nlemfuensis* is the more important weed in East Africa, but one or other species is common from sea level to 2000 m. *Cynodon* does not constitute as serious a weed problem in East Africa as *Digitaria abyssinia*, but is troublesome in a wide range of plantation and arable crops, especially coffee.

Methods of control
C. dactylon is difficult to control by cultivation because of its underground rhizomes, and in such perennial crops as coffee the amount of cultivation required is liable to damage the crop roots. It is susceptible to dalapon, however; doses of 5—10 kg per ha applied to young, actively growing grass, preferably at the beginning of the rainy season, giving a high degree of control. A single application rarely gives complete eradication, but spot treatment is generally sufficient to deal with regrowth. Coffee bushes can be injured by excessive doses of dalapon and 6 kg per ha is the maximum normally recommended in coffee. Doses of the same order can also be employed in a number of other plantation crops. Most annual crops are damaged by dalapon, so that the weed must be eradicated prior to planting.

Glyphosate has the advantage over dalapon of being inactivated on

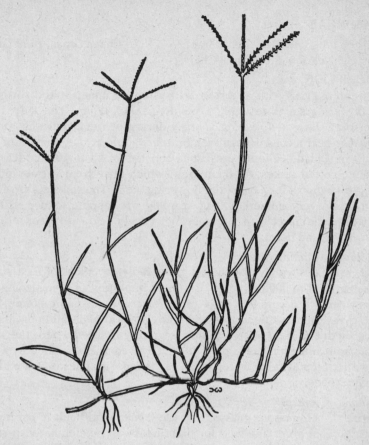

Figure 17. Cynodon nlemfuensis

contact with the soil and applied as a directed spray at rates of about 3 kg per ha is effective and safer in some plantation crops. Amino-tria-zole is effective against *C. dactylon* in similar situations to dalapon. Paraquat kills back the shoots and, by repeated application, can be used to maintain control through the growing season.

C. nlemfuensis, lacking rhizomes, is presumably considerably easier to control by mechanical or chemical methods than *C. dactylon*. Con-fusion of the two species may account for some of the variation in res-ponse to treatment which has occurred in the past.

GRAMINEAE

Digitaria abyssinica (A. Rich.) Stapf   Couch Grass, Blue Couch
   (= *D. scalarum* (Schweinf.) Chiov.)

DESCRIPTION

A perennial grass with an extensive system of underground rhizomes and flowering stems reaching a height of 30—45 cm. The leaves are softer than those of *Cynodon,* usually dark green in colour and, unlike *Cynodon,* bear a conspicuous membranous ligule at the junction of leaf sheath and blade. The inflorescence consists of up to 10 slender, upwardly directed spikes, 6—8 cm long, which arise from various points along the topmost 2—7 cm of the flowering stem. The spikelets are about 2 mm long, very numerous, but scarcely overlapping, and consist of a single fertile floret with a membranous scale below it representing a reduced, second floret.

Distribution and importance

D. abyssinica is widely distributed in the moister regions of East Africa from sea level to 3000 m and is the most important of the rhizomatous grass weeds. It occurs in a wide range of crops, including coffee, tea, sisal, pyrethrum, wattle, cotton and many other annuals and perennials. The growth and yield of crop plants is greatly reduced where the weed occurs and coffee bushes in particular can be completely killed by a bad infestation. It may fairly be regarded as the most troublesome of all East African weeds.

Methods of control

As with all rhizomatous grasses, *D. abyssinica* is difficult to control by cultivation. In preparing land for planting it can be eliminated if sufficient cultivations are carried out but, once established in a crop, it is rarely possible to remove all the rhizomes even by the expensive process of forking out. In long-term crops it is especially difficult to remove as the rhizomes penetrate among the crop roots, where considerable damage can be done by attemps to dig out the weed.

    Digitaria is, however, relatively susceptible to the effects of dalapon at doses of 5—10 kg per ha. The best results are obtained by spraying at medium to high volume (500—1000 l per ha) at a time when the young grass shoots are growing rapidly. The start of the rainy season is probably the best time, but fresh growth can be stimulated later in the season by cutting. In some trials good results have also been obtained by applying the chemical to the soil when new shoots are just starting to emerge after cultivation but more experience of this technique is needed before it can be recommended. Although a single application of dalapon

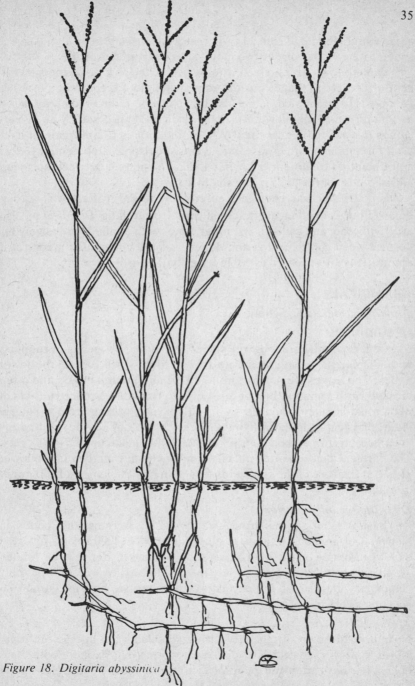

Figure 18. Digitaria abyssinica

gives a high degree of control, it is usually necessary to apply a second spot treatment after one or two months to deal with regrowth.

Dalapon can be used in well established coffee as long as the spray is kept off the crop foliage, but the rate should not exceed 6 kg per ha if damage is to be avoided. It can also be applied as a directed spray in tea and mature sisal, and low doses have been used successfully as overall sprays in sisal bulbil nurseries. Where *Digitaria* is troublesome in land used for annual or other susceptible crops, dalapon can be employed as a pre-planting treatment, allowing a suitable interval between spraying and planting for residues in the soil to disperse.

Amino-triazole has been less extensively tested in East Africa than dalapon but is another useful chemical for controlling *Digitaria* as also are glyphosate and asulam. Paraquat, now being applied extensively for general weed control in tea and coffee, will only give lasting control if applications are repeated every few weeks.

GRAMINEAE

Diplachne caudata K, Schum.

DESCRIPTION
A tufted, perennial grass of wet places, spreading by means of runners, with slender, flowering stems up to 1 m high which root at the lower nodes. The leaves are narrow, up to 30 cm long, rough-edged and often inrolled, with a short (1 mm or less) ligule. The inflorescence consists of numerous, slender branches up to 10 cm long arising from a central stem at an acute angle. Each branch bears numerous sessile, alternately arranged, narrow spikelets, 6—9 mm long and made up of 3—5 florets. The florets, which are often purplish, only overlap slightly and the lower glume at the base of the spikelet is less than half the length of the upper glume.

Distribution and importance
An East African species from swampy land, streams and ponds in Kenya, Tanzania and Uganda at altitudes of 300—2000 m and reported as a weed of rice on the Mwea irrigation scheme in Kenya. The related *D. fusca* is distributed more widely in Asia and Africa, where it not only competes with the crop but also acts as an alternate host for certain insect pests of rice.

Methods of control
No information on control measures is available for this species. In South Korea, however, control of *D. fusca* is reported by treatment with butachlor a few days after transplanting rice seedlings and, being of a similar

habit of growth, *D. caudata* would be expected to react in a similar manner.

Figure 19. Diplachine caudata

GRAMINEAE

Echinochloa colona (L.) Link   Barnyard Grass

DESCRIPTION

A coarse, annual grass usually 30 —75 cm tall, with an inflorescence of 8 —10 short, densely crowded spikes at the top of the stem. The stems usually grow outwards at the base before turning upwards and are often purple near the base. The leaves are hairless, up to 25 cm long and 3 —8 mm wide, sometimes banded with purple and without a ligule. The spikes making up the inflorescence are up to 3 cm long and 3 —4 mm wide, usually about half their length apart on the main stem, which they join at an acute angle. They are made up of numerous almost stalkless spikelets arranged in four distinct rows. Individual spikelets are 2 —3 cm long, oval in shape with a pointed tip (but not extended into an awn as in the related *E. crus-galli*) and contain a single fertile floret.

Distribution and importance

An important crop weed throughout the tropics and subtropics, *E. colona* occurs widely in East Africa from sea level up to 2000 m. It grows very vigorously on wet land and is strongly competitive. In rice, a crop which it somewhat resembles in the seedling stage, it is often the dominant weed. It has also been recorded as a weed in cotton in Kenya and can occur in other crops, especially under irrigated conditions.

Methods of control

Care in the preparation of rice ground for planting and in the manipulation of water levels can minimize the problems caused by grass weeds. If problems of grass weeds develop, yields will inevitably suffer unless the crop is hand-weeded, a laborious method, or a herbicide treatment applied. Although there is only limited information on the chemical control of *E. colona* in East Africa, the results available support evidence from other rice-growing areas that propanil can be used successfully as a post-emergence treatment. This herbicide must be applied while the weed is still young (2 —4 leaves) and the ground should be flooded 12 —24 hours after spraying. A number of newer chemicals, including butachlor and oxadiazon, either used alone or in combination with propanil also appear to be effective against this weed.

Figure 20. Echinochloa colona

GRAMINEAE

Eleusine indica (L.) Gaertn.   Wild Finger-millet, Crows-foot Grass

Subspp. *indica* and *africana* (Kennedy O'Bourne) S.M. Phillips

E. multiflora A. Rich.

DESCRIPTION

Both species are wiry-rooted, annual grasses growing up to 50 cm high, with flattened shoots, leaves folded at the base and an inflorescence of several spikes clustered at the top of the stem.

In *E. indica* there are most commonly 3 —8 narrow spikes 5 —10 cm long, about 5 mm wide, which radiate from the stem apex, one of them often arising shortly below the apex. The numerous sessile spikelets are about 5 mm long and densely crowded into two rows on the lower side of the spike. Each spikelet consists of 3 —9 florets, without bristles (a feature distinguishing this species from *Chloris pycnothrix* and *Dactyloctenium aegyptium* which have a similar inflorescence structure).

The two subspecies differ mainly in that subsp. *africana* is a tetraploid and more robust than the diploid subsp *indica*. In subsp. *africana* the florets are 3.7 —4.9 mm long and the ligule, where the leaf blade meets the stem, has a fringe of hairs, whereas in subsp. *indica* the florets are 2.4 —3.6 mm long and the ligule is not obviously hairy.

E. multiflora is similar in its general habit of growth but the greyish spikes making up the inflorescence are short and broad, not more than 3 cm long and arise alternately over a short section at the top of the stem. The spikelets are up to 10 mm long and contain up to 15 florets which are abruptly narrowed to a short point at the tip.

Distribution and importance

E. indica is one of the most widespread and troublesome annual grass weeds in tropical and sub-tropical parts of the world and is common throughout East Africa at altitudes up to 2000 m. Subsp. *africana* occurs most commonly above 300 m in eastern and southern Africa. Both types are important as weeds of waste ground and crops, including cotton, maize and sugar cane. *E. multiflora* is an eastern African species found above 1500 m in Kenya and Tanzania. It often appears on disturbed ground near stock camps and as a crop weed, especially in the Kenya Rift Valley.

Methods of control

When just emerging from the soil *E. indica* seedlings can be controlled with 2,4-D and this herbicide can be used in maize as a pre-emergence treatment. Among the residual chemicals it is somewhat resistant to tri-

Figure 21. *Eleusine indica*

azines, including atrazine, but susceptible to such chemicals as alachlor, metolachlor, EPTC, trifluralin etc. applied as pre-emergence treatments. A combination of atrazine and alachlor is often used in maize. Once the plants are established they are more difficult to control. Glyphosate can be used in situations where it can be directed away from the crop and asulam gives control in sugar cane.

There is no indication that *E. multiflora* differs appreciably from *E. indica* in its response to herbicides.

GRAMINEAE

Eleusine jaegeri Pilg.   Manyatta Grass

DESCRIPTION

A coarse, tufted perennial forming dense tussocks, up to 1 m high and as much across. The shoots and leaf sheaths are laterally flattened and the leaves stiff, up to 60 cm long with sharp edges. The inflorescence consists of 3 —7 stiff, one-sided, greyish green spikes, up to 15 cm long, arising from the top 5 —8 cm of the stem. The spikelets are numerous and densely crowded along the underside of the spike. They are 5 mm long and consist of 4 —5 florets.

Distribution and importance

E. jaegeri is a grass of high altitudes in East Africa, occurring widely at altitudes above 2000 m in Kenya, Tanzania and Uganda. It is an undesirable plant in grazing land as it is avoided by stock and the common name is derived from its frequent occurrence on the sites of abandoned Masai cattle enclosures. It often occurs in association with *Pennisetum sphacelatum* (Nees) Dur. & Schinz (= *P. schimperi* A. Rich.) (Wire Grass), another unpalatable, tufted species (distinguished by its bristly, unbranched, cylindrical flower heads) and both species are frequent invaders of highland pastures declining in fertility or of *Themeda triandra* (Red-oat Grass) grassland where grazing is too intensive to permit occasional burning.

Methods of control

Invasion of *E. jaegeri* into *Themeda* grassland can be prevented by periodic controlled burning and in Kikuyu grass pasture its spread can be discouraged by the maintenance of a high level of fertility. Where *E. jaegeri* and *P. sphacelatum* are the dominant grasses, however, they are largely resistant to fire and their removal is difficult. Spot treatment of young growth with dalapon at 10 —15 kg per ha may be a possible treatment for small areas and glyphosate could also be used in this way. For larger areas overall treatment with dalapon at 4 —6 kg per ha is

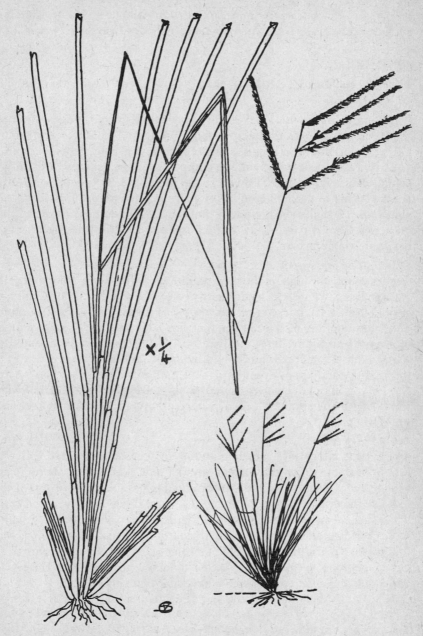

$\times \frac{1}{4}$

Figure 22. Eleusine jaegeri

reported to check *Eleusine* and *Pennisetum* without seriously affecting such desirable species as *Chloris gayana* (Rhodes Grass) or *Themeda*.

GRAMINEAE

Imperata cylindrica (L.) Raeuschel   Swordgrass, Lalang

DESCRIPTION
A perennial grass with an extensive and deeply penetrating system of rhizomes, and shoots reaching a height of 1 m. The leaves are rather stiff and upright, with rough, sharp edges and a long, tapering point. The inflorescence is branched, but compacted into a dense, white, fluffy, cylindrical, spike-like head up to 15 cm long by 2 cm broad. The fluffy nature is due to the numerous silky hairs arising from the base of the spikelets. The spikelets are 5 mm long and arranged densely along the branches. They consist of one fertile floret and a sterile floret reduced to a pair of small bracts.

Distribution and importance
Occurs widely throughout the tropics, in East Africa being commonest in coastal areas, where it is a serious weed of coconut plantations and occasionally of sisal. It also occurs as a weed of roadsides and waste ground in a number of the moister inland districts including the Nyanza district of Kenya and many parts of Uganda. In the Southern Highlands of Tanzania it is sometimes a problem in coffee, and is also recorded as a weed in tea and various other crops. Although sometimes a bad weed, *Imperata* is of relatively minor importance in East Africa compared with West Africa and south-east Asia, where it is the worst of the perennial grass weeds.

Methods of control
Imperata is intolerant of shading and in such crops as tea and coffee is less important where shade trees are present than where there is no shade. It is difficult to eradicate by cultivation because of its rhizome system but thorough cultivation repeated several times can give good control, especially when done under conditions where the rhizomes brought to the surface can dry out. In parts of south-east Asia it has been found that intensive grazing by cattle gives some measure of control and various cover crops, such as *Pueraria phaseoloides* var. *javanica, Tephrosia candida* and *Euphorbia geniculata* have also been effective.

In Malaysia dalapon at 10—15 kg per ha gives a considerable degree of control. It is most effective on young growth about 50 cm high but a follow-up spray or cultivation is usually needed to prevent regrowth. Glyphosate at 2—4 kg per ha is also an effective chemical. Paraquat kills the

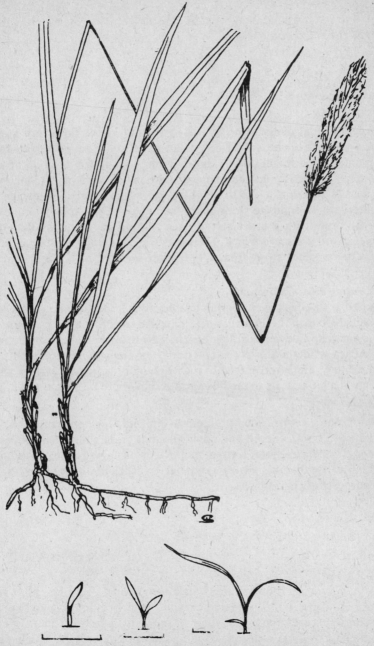

Figure 23. Imperata cylindrica

top growth but the treatment must be repeated at intervals as it has little effect on the rhizome system.

GRAMINEAE

Lolium temulentum L.   Darnel

DESCRIPTION

An erect, annual grass growing 0.3 —1 m high, with flowering stems which are rough towards the tip. The leaves also are rough, up to 1 cm broad and have a membranous ligule 2 mm long at the junction of leaf blade and sheath. The inflorescence is unbranched and about 15 cm long with sessile spikelets 1 cm long which are borne singly, edgeways on and alternate on opposite sides of the main axis. Each spikelet consists of 4 —5 florets with a bract at the base as long as or longer than the spikelet itself; a feature which distinguishes *L. temulentum* from cultivated *Lolium* species (ryegrass). The outer bract of each floret bears a long bristle at its tip.

Distribution and importance

Recorded as an occasional weed in arable crops and waste land between 1800 and 2200 m in the Eldoret and Nakuru areas of Kenya, *L. temulentum* is native to Europe and Asia and has probably been introduced into East Africa with seed of cereals. The seeds are very objectionable in grain as they are often infected with a fungus which imparts a grey colour and a bitter taste to the flour and is thought to be poisonous.

Methods of control

Although not a very serious problem at present, further spread of this weed should be prevented and contaminated grain should not be used for seed. In trials at Njoro, Kenya, metoxuron has given control as a post-emergence treatment in wheat and barley. Diclofop-methyl has been used successfully in Canada.

GRAMINEAE

Oryza punctata Steud.

O. barthii A. Chev   Annual Wild Rices

DESCRIPTION

Oryza punctata is a robust, tufted, hairless, annual grass reaching 1 m or more in height, with a loosely branched panicle of long-awned spikelets. The stems are spongy and 4 mm or more in diameter. The leaves up to 30 cm long × 1 cm wide, tapering to a sharp tip, with two small auricles where the blade joins the sheath and a soft ligule 3 —10 mm long.

Figure 24. Lolium temulentum

Figure 25. Oryza punctata

The inflorescence branches are spreading and up to 20 cm long; the spikelets borne singly on short stalks on the main and side branches, usually overlapping slightly. Individual spikelets are 5—6 mm long, laterally compressed and consist of two hard scales, the outer one ending in an awn up to 7 cm long.

O. *barthii* is similar in general appearance but differs in its denser, more upright inflorescence and in its larger spikelets, 7—11 mm long with an oblique attachment to the stalk. The awns are also longer, up to 16 cm.

Distribution and importance

O. *punctata* is largely an African species, also found in Madagascar and Thailand, which grows in wet ground alongside streams and ponds from sea level to 1200 m. In East Africa it is recorded from coastal areas, including Zanzibar, in northern and western Uganda and in various parts of Tanzania. It readily becomes established as a weed in rice paddies and is noted as important in this crop at Mbarali in southern Tanzania, where populations build up to problem levels after three years of cropping.

O. *barthii* is more of a problem in West than East Africa but occurs in central and southern Tanzania, growing in shallow water and sometimes becoming a weed in rice.

Methods of control

The build-up of O. *punctata* populations can be prevented by rotating rice with broad-leaved crops in which control of grass weeds is easier. In soyabeans, for example, such chemicals as alachlor or metolachlor can be used effectively as pre-emergence treatments to control this species. Selective control in rice is more difficult because crop and weed are so closely related. Under experimental conditions there are indications that it may be possible to use alachlor and certain other chemicals in the rice crop itself through use of the safener naphthalic anhyride, applied as a seed dressing. In Senegal dalapon has been used to control O. *barthii* as a fallow treatment and oxadiazon has also been reported promising.

GRAMINEAE

Panicum trichocladum K. Schum.

DESCRIPTION

A trailing, perennial grass with shoots up to 3 m long, rooting at the nodes or scrambling over taller vegetation. The leaves are up to 15 cm long, more or less softly hairy and with a whitish midrib. The upper leaf sheaths clasp the stem tightly, the lower ones are loose and often come

Figure 26. Panicum trichocladum

away from the stem. The ligule at the mouth of the leaf sheath is inconspicuous. The inflorescence is erect and loosely branched, up to 15 cm long × 8 cm across, the branches very slender and covered with long, white hairs. The stalks of the spikelets are often paired, the pairs being of very unequal length. The spikelets themselves are pale green or flushed with purple, about 3 mm long, and made up of two florets, only the upper one fertile. The two bracts enclosing the spikelet are very dissimilar, the inner one equal in length to the lower floret, the outer much smaller.

Distribution and importance

P. trichocladum is an indigenous species, widely distributed in Kenya, Tanzania and Uganda from the coast to about 1800 m. It is a common weed of arable crops and waste land, particularly in coffee, where it has been reported from all the main growing areas. It is very common in sisal, especially in coastal districts, and, although in field sisal it appears to have little effect on the growth of the crop, it can be very objectionable in bulbil nurseries.

Methods of control

As its runners are all above ground, this grass is not difficult to control by cultivation. When it becomes established in a mulched crop, however, as often happens with coffee, mechanical cultivation is not possible. Relatively little is known about the effect of herbicides and trials in East Africa have given somewhat conflicting results. Control has been achieved with TCA, which may be of value in uncropped land, but cannot be used in coffee. Where dalapon has been used to control *Digitaria scalarum* in coffee an increase in *P. trichocladum* has sometimes been observed. In sisal, on the other hand, dalapon has been an effective treatment and the increase observed in coffee may have been due to subsequent invasion rather than to survival of spraying. Among the soil-acting herbicides, monuron at a dose of 8 kg per ha has been reported to give control in sisal in Tanzania, as has bromacil at 2 – 3 kg per ha.

GRAMINEAE

Paspalum scrobiculatum L.

(= *P. commersonii* Lam.)

DESCRIPTION

A coarse, tufted, annual or short-lived perennial grass growing to a height of 60 cm or more, with an inflorescence of 1 – 4 (sometimes more) upwardly directed, flattened spikes near the top of the stem. The leaves are usually 10 – 25 cm long × 5 – 10 mm wide, tapering gradually to a

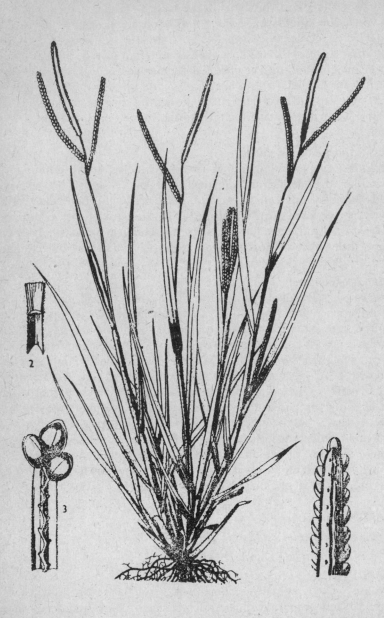

Figure 27. Paspalum scrobiculatum

fine tip and with a membranous ligule not more than 1 mm long. The degree of hairiness varies greatly. The spikes arise singly from the top 5 cm or so of the flowering stem and are mostly 5 —8 cm long × 2 —3 mm wide, consisting of a flattened axis with two rows of flattened, circular spikelets on the underside. The spikelets are about 2 mm in diameter, with a single fertile floret enclosed between two scales.

Distribution and Importance

A species common in tropical Asia and Africa, *P. scrobiculatum* is found throughout East Africa from sea level to 2900 m. It usually occurs as a weedy species associated with cultivation or pastoral farming and is especially prevalent in moist, shady situations. In pasture land it is a useful grazing grass, but a highly competitive weed in crops such as maize, groundnuts and sugar-cane and a common weed of roadsides and waste ground.

Methods of control

There is little specific information about the control of this species in East Africa but in sugar cane in South Nyanza mixtures containing atrazine and alachlor have not given good control and similar mixtures are reported as only partially effective in maize in West Africa. In several countries glyphosate has been found to give appreciably better results than paraquat or dalapon.

GRAMINEAE

Pennisetum clandestinum Chiov.   Kikuyu Grass

DESCRIPTION

A creeping, mat-forming, perennial grass, with stout rhizomes below the surface and long runners above ground, which produce tough, fibrous roots and short, stout, vertical branches from the nodes. The internodes are short and the leaves are up to 30 cm long. Unlike other common grasses, the flowers are not borne at the top of a long stem, but are enclosed within the topmost leaf sheaths of short side shoots, similar in general appearance to non-flowering shoots. The only parts of the flower visible are the stamens, which appear as masses of fine white threads and consist of filaments 2 cm or more long, with anthers at the tip. The spikelets themselves are about 1 cm long and borne in groups of 2 —4 in the upper leaf sheaths. They are surrounded by a number of slender bristles and each contains a single fertile floret. Seeds are produced, but not in large numbers and, as it is almost impossible to harvest, establishment of Kikuyu grass pastures is normally by planting stem cuttings.

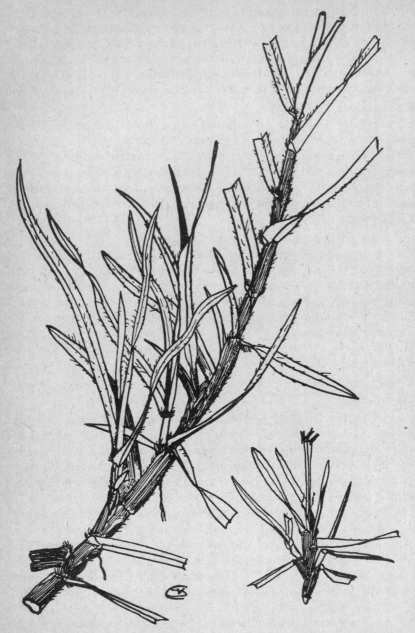

Figure 28. Pennisetum clandestinum

Distribution and importance
Kikuyu grass is indigenous to the highlands of East Africa and was originally of limited distribution. It has been widely planted, however, both as a pasture and a lawn grass and grows well throughout East Africa at altitudes above about 1500 m where the annual rainfall is 1000 mm or more. It has also been planted in other parts of the world including the U.S.A. and New Zealand. In its natural habitat its dominance is dependent on a high level of fertility and it is replaced by less desirable grasses such as *Eleusine jaegeri* and *Pennisetum sphacelatum* (= *P. schimperi*) under conditions of soil deterioration.

Kikuyu grass has been recorded as a weed in pyrethrum, tea, coffee and many arable crops and is difficult to eradicate by cultivation unless weather conditions are dry for a prolonged period. It can be particularly troublesome when pyrethrum is grown after a Kikuyu grass ley.

Methods of control
Early studies with dalapon showed that application to rapidly growing *P. clandestinum* gave a large measure of control but that further treatment was generally needed to deal with regrowth. More recently, glyphosate has proved effective on this species and a rate of 2 kg per ha usually gives good results. Glyphosate is not a selective treatment but its lack of residual activity in the soil makes it possible to plant crops soon after treatment. Direct drilling into glyphosate treated grass is also a possibility with certain crops.

GRAMINEAE

Rottboellia cochinchinensis (Lour.) Clayton   Guinea-fowl Grass, Itch
  (= *R. exaltata* L.f.)   Grass

DESCRIPTION
An annual grass of vigorous growth, often reaching 2.5 —3 m in height, the leaf sheaths bearing long, brittle hairs, which cause irritation if they come in contact with the skin. The leaves are 60 × 2.5 cm wide, with well marked, white midribs and rough, sharp edges. The inflorescence consists of a number of unbranched, narrow, cylindrical spikes 8 —15 cm long, arising in the axils of the upper stem leaves, their stalks largely enclosed by the somewhat inflated leaf sheaths. The main axis of the spike is thickened and jointed, each joint bearing a pair of small spikelets, one sessile, the other on a short stalk. When ripe, the spike breaks at the joints into hard, cylindrical sections about 5 mm long.

Distribution and importance
A grass of widespread distribution in the tropics and sub-tropics which

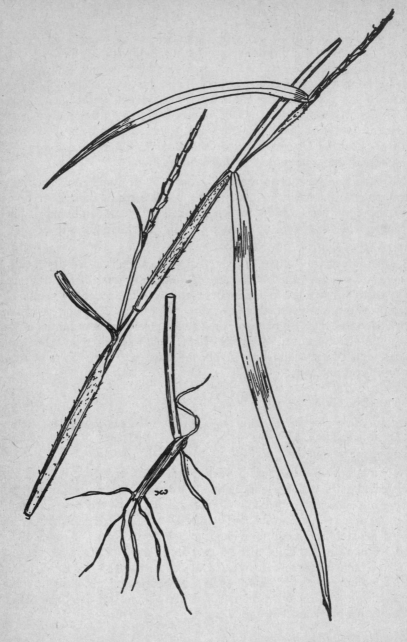

Figure 29. Rottboellia cochinchinensis

is common throughout East Africa from sea level to 2000 m. It is one of the primary colonizers of disturbed ground and may be of some value for grazing. It can be a troublesome weed in maize, rice, sugar cane and other arable crops, however, where its rapid growth and spreading habit make it a very competitive plant.

Methods of control
Seeds of *Rottboellia* have only a short dormancy period and control is not difficult where cultivation is possible. Once established, however, the plants grow very rapidly and, although inter-row cultivation deals with weeds growing between the rows, those within the rows can still be troublesome. Among foliar acting herbicides, diclofop-methyl can be used for control in various broad-leaved crops, propanil has given good results in rice and asulam is an effective treatment in sugar cane. Among the residual herbicides diuron has been found effective as a pre-emergence treatment in sisal, trifluralin (incorporated pre-planting) in cotton, groundnut and soyabean and pendimethalin (also soil incorporated) appears promising in maize.

GRAMINEAE

Setaria verticillata (L.) Beauv.   Bristly Foxtail, Love Grass
S. pumila (Poir.) Roem & Schult.   Pale Foxtail
   (= *S. pallide-fusca* (Schum.) Stapf & Hubbard)

DESCRIPTION
Setaria verticillata is a tufted annual grass, most readily recognized by the bristly, cylindrical inflorescence which readily breaks off and sticks firmly to clothing. It grows up to 60 cm tall, with the shoots spreading at the base and rooting from the nodes. The leaves are up to 15 cm long × 12 mm wide, dark green, soft in texture and with well marked lateral veins. The inflorescence is up to 8 cm long × 15 cm diameter, often tinged with purple and made up of numerous clusters of spikelets arising in whorls from the main axis. The spikelets are shortly stalked and consist of a single fertile floret with a reduced, sterile floret bearing a stiff, barbed bristle, 5 mm or more long arising from the base.

S. pumila is similar in growth habit and differs most clearly in each spikelet bearing 6 —8 bristles and in the bristles having teeth directed forward instead of backwards, so that they are less adhesive. Other differences include a denser, less obviously whorled inflorescence and in the presence of scattered, long hairs at the base of the leaf blade.

Distribution and importance
Both species are important as weeds in many parts of the world, both in

Figure 30. Setaria verticillata

the tropics and in warmer temperate regions. In East Africa they are common and widespread in arable and waste land at all altitudes, often becoming very conspicuous in areas newly cleared of bush. They are among the commonest annual grass weeds in coffee and also occur in most other crops. In grass leys *S. verticillata* is often prominent in the early stages of establishment and, although a useful grass for grazing, the flower heads can be troublesome through adhering to the skin of stock.

Methods of control

Mechanical control presents no special problems and chemicals are available which enable the annual *Setaria* species to be controlled effectively in a variety of crops. The following list (not intended to be complete) indicates the range of possibilities: trifluralin can be used as a pre-sowing treatment in cotton and beans, alachlor or metolachlor pre-emergence in maize, pendimethalin pre- or metoxuron early post-emergence in wheat, metribuzin pre- or post-emergence in tomato and asulam post-emergence in sugar cane.

GRAMINEAE

Snowdenia polystachya (Fresen) Pilg.   Abyssinian Grass

DESCRIPTION

A tufted annual or short-lived perennial grass growing 1.5 m or more high with an elongated, leafy inflorescence of numerous bristly spikes. The stems are often 4 –5 mm in diameter at the base, spreading and rooting at the lower nodes, erect or scrambling above. The leaf blades are up to 30 cm long × 10 mm or more wide, the sheaths often tinged with pink at the base and more or less inflated above. Near the top of the sheath is an area of long, bulbous-based hairs, also found sparsely on the blade, and the short ligule is fringed with long hairs. The 2 –5 cm long flower spikes develop in the upper part of the main stem and side branches on stalks about the same length as the leaf sheaths, so that they project in the angle between the blade and the stem. The shortly stalked spikelets are 2 –3.3 mm long, densely crowded and consist of a single fertile floret enclosed by two scales, the outer one with an awn up to 7 mm long and often pink in colour at the tip.

Distribution and importance

S. polystachya is a grass of north-eastern Africa which is thought to have been introduced into Kenya as a forage crop. It is now established as a weed in a number of localities in the Nakuru and Kiambu districts of Kenya and in northern Tanzania at altitudes of 2200 –2700 m and ap-

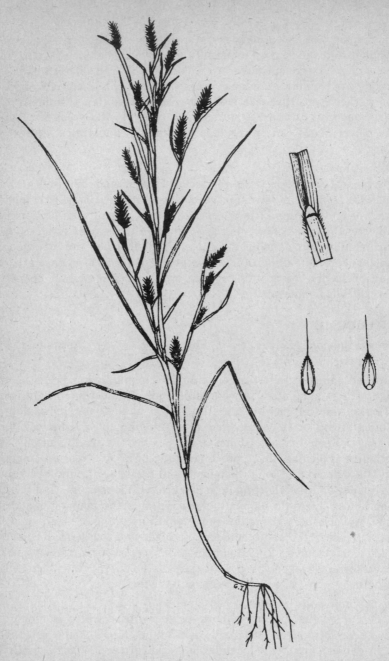

Figure 31. Snowdenia polystachya

pears to be spreading. It has become an important weed of wheat in Kenya and is likely to cause problems in other crops of higher altitudes.

Methods of control
Trials at Njoro have shown that pre-emergence treatment with alachlor or propachlor controls this species and can be used selectively in maize and beans. In trials in Ethiopia diclofop-methyl has given good control in wheat and barley and, although specific information on other herbicides is lacking, on the basis of their effectiveness against other annual grasses, pendimethalin and metoxuron would also be expected to be possible treatments in wheat.

TYPHACEAE

Typha domingensis Pers.   Reedmace
  (= *T. angustifolia* auct. non L.   Bulrush
  and *T. australis* Schumach. & Thonn.)   Cat's Tail

T. latifolia L.

T. capensis (Rohrb.) N.E.Br.

DESCRIPTION
Typha species are tall, perennial herbs growing in swamps and shallow water to a height of 2 – 3 m or more, with erect, unbranched stems growing from a stout rhizome and long, grass-like leaves. Individual flowers are minute, but very numerous (up to 220,000 female flowers have been estimated to be present in a single head of *T. latifolia*) and are densely clustered round the swollen, terminal part of the stem to form a long, cylindrical spike. The upper part of the spike consists of male flowers which fall off, leaving a bare stem, as the lower, female portion ripens. The flowers are much reduced, consisting either of a group of 2 – 4 stamens or a stalked ovary surrounded by fine hairs and elongated bracts. The species are all similar in general appearance, but can be distinguished as follows:—

T. domingensis; male and female parts of spike separated by a 1 – 2 cm gap, male part as long as or longer than female, which is reddish brown in colour.

T. latifolia; male and female parts of spike not separated by a gap, male part normally shorter than female. Lower part of spike 2.5 – 3.5 cm wide, yellow green when young, later becoming dark brown.

T. capensis; similar to *T. latifolia,* but leaves often broader (up to 15 mm compared with 10 mm). Female parts of spike narrower, 1.5 – 2 cm, and reddish brown at maturity.

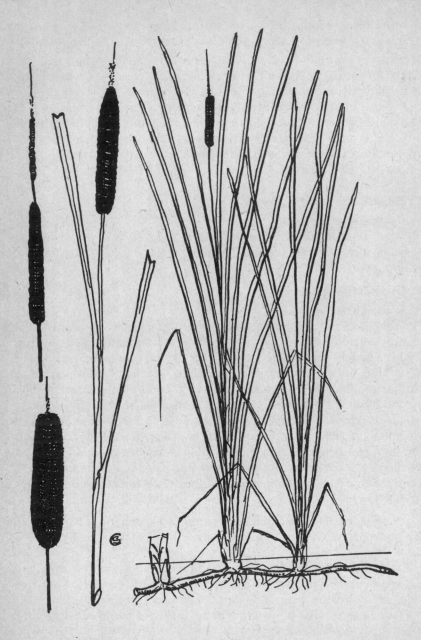

Figure 32. Typha spp.

Distribution and importance
All three species occur in the same type of habitat at the edges of
swamps, dams, lakes and rivers. *T. domingensis* is the commonest in
East Africa and is found in suitable habitats throughout the region.
Above 1600 m in Kenya it is frequently associated with *T. latifolia*. The
other species are more restricted in distribution. *T. capensis* is a southern
African type recorded in Tanzania and Uganda, but not in Kenya. *T.
latifolia* on the other hand has a north temperate distribution and extends
south as far as Kenya and Uganda without having been found in Tanzania. The plants are important as weeds when they become established
in dams, reservoirs, drainage ditches, etc., in which situations they
interfere with the flow of water and encourage silting.

Methods of control
Typha species generally can only grow in shallow water and their establishment can be prevented if the depth of water in the dam or re-
servoir is maintained at more than 1.2 m and precautions are taken
against silting. It may sometimes be possible to kill an existing infestation by increasing the depth of the water, but in the majority of cases
this is not possible and some other means of eradication must be adopted. Relatively little experience of the use of herbicides against *Typha* is
available in East Africa but in Britain, America and elsewhere a number
of materials have given satisfactory control.

Relatively high doses of MCPA or 2,4-D (5–10 kg per ha) applied at
flowering time will control *T. latifolia*, but dalapon and amino-triazole
are now in more general use for this purpose. Dalapon is generally applied at about 20 kg per ha, amino-triazole at about 10 kg per ha and
mixtures of the two have also been effective. Care must be taken to ensure that water used for irrigation or drinking is not contaminated, but
both dalapon and amino-triazole are harmless to fish at the doses applied
for control of aquatic weeds. More recently it has been shown that glyphosate gives good control at 2–3 kg per ha and this chemical does not
present residue problems.

WOODY WEEDS

CACTACEAE

Opuntia species   Prickly Pear, Cactus

DESCRIPTION

Shrubby cacti growing up to 3 m high, with spiny or occasionally spine-less branches and large yellow flowers. The stems are made up of thick, flat and oval or cylindrical green joints (up to 30 cm long × 15 cm wide) and are leafless, except at the tips of the youngest shoots. The whitish spines are straight and sharp, 2 —4 cm long and arranged in groups of 2 or 3 in a spiral pattern round the stem. There are also clusters of nume-rous spiny hairs which very easily break off and become embedded in the skin if the plant is touched. The flowers are up to 8 cm across and borne singly on the upper edges of the young joints. The petals are nu-merous and there are a large number of stamens, the fruit is fleshy and pear-shaped, 4 —5 cm long and can be eaten after the spiny skin has been peeled off.

At least four species of *Opuntia* occur as weeds in East Africa. *O. vul-garis* Mill. (formerly misidentified as *O. dillenii*) is common along the coast and also inland, *O. ficus-indica* (L.) Mill, which is usually more or less spineless, *O. exaltata* A. Berger, which has cylindrical spiny stem segments, and an unidentified species also occur inland.

Distribution and importance

Originating in America, prickly pear species have been introduced into many tropical and sub-tropical countries and are now widely distributed. Plants in East Africa have probably been introduced from several differ-ent sources. The thornless type is a valuable dry season fodder plant for cattle, but the thorny type is more widespread, having often been plant-ed as a hedge. It is common in coastal areas of Kenya and Tanzania and in Zanzibar. There are also bad infestations in the Rift Valley and Nyanza districts of Kenya and along the eastern shore of Lake Victoria. Small patches occur in many other districts.

Once a prickly pear hedge is established, it is very difficult to control as cut or broken fragments of stem which fall to the ground readily take root. Hedges thus become steadily thicker and in a few years consider-able areas of dense thicket develop which can neutralize large areas of pasture land.

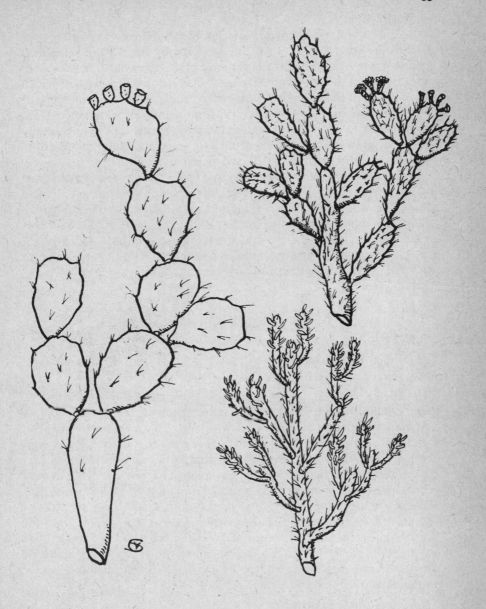

Figure 33. Opuntia spp.

Methods of control

Biological control has been very successful against some of the *Opuntia* species, the control of *O. stricta* in Australia by the moth *Cactoblastis cactorum* providing the classical example of the potential of this method. Various species of the cochineal insect, *Dactylopius*, have also provided a useful degree of control in a number of countries and introductions of *D. tomentosus* have been made into Kenya and Tanzania since 1958. Severe injury to *Opuntia* has resulted in some areas, especially near the coast, but the spread of the insect is very slow and it is probable that periodic redistribution will be needed to maintain effectiveness.

Chemical control can also be effective and before its use was banned in East Africa, 2,4,5-T gave good results when applied in an oil-based spray. The alternative brush-killer picloram has been found effective against *Opuntia* in the U.S.A., applied either as a spray or in the form of pellets. The picloram/2,4-D formulation marketed in East Africa would also be expected to have a potential for control of this weed. In South Africa MSMA shows promise against *Opuntia*.

CAESALPINIACEAE

Caesalpinia decapetala (Roth) Alston   Mauritius Thorn
(= *C. sepiaria* Roxb.)

DESCRIPTION

A scrambling, woody climber with large heads of pale yellow flowers and stems up to 6 m long covered with very sharp, recurved prickles. The alternately arranged leaves are doubly pinnate and up to 30 cm long, the main axis of the leaf also bearing sharp prickles on the underside. The pinnae are arranged in opposite pairs and are divided into numerous pairs of oblong leaflets up to 25 mm long × 12 mm wide. The flowers are borne on shortly hairy stalks up to 25 mm long and are arranged to form terminal and lateral racemes up to 30 cm long. They are about 1 cm across and made up of 5 sepals, 5 petals, the upper ones smaller than the lower, 10 long, hairy stamens arranged in a cylindrical, downward pointing group and an ovary which develops into a woody flattened, dark brown pod about 8 cm long.

Distribution and importance

Mauritius Thorn appears to have been introduced from India, via Mauritius, and has been widely planted as a hedge plant in many of the wetter districts of East Africa at altitudes of 1000—2000 m. It is particularly

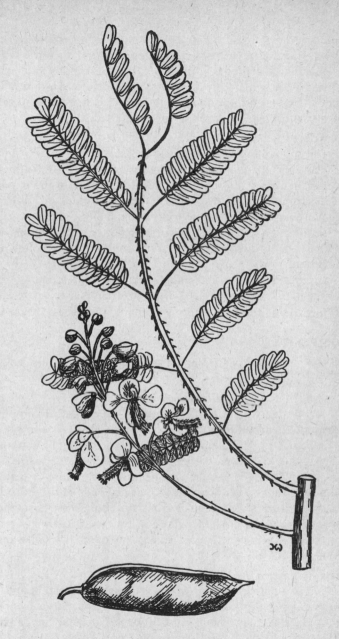

Figure 34. Caesalpinia decapetala

common in the Arusha district of Tanzania and the Kericho district of Kenya. It makes a very effective hedge, but requires a lot of room and, unless cut back at frequent intervals, spreads rapidly and forms impenetrable thickets which can be very troublesome in grazing land. It grows readily from seed and, once planted as a hedge, it soon becomes established as a weed on waste ground and roadsides nearby.

Methods of control

The arching growth of the stems makes it very difficult to cut the base of the plant and manual clearing is very difficult. Formulations of brush-killer chemicals in diesel oil applied to the base of the stems or water-based sprays on the foliage will kill the parts contacted. With large thickets, however, it is difficult to treat enough of the plant to obtain a complete kill, so that regrowth occurs. Pelletted formulations of picloram which can be thrown by hand into the centre of the thickets have proved effective on this species.

Mauritius Thorn thickets usually contain large amounts of dead wood and are highly inflammable so that, where it is safe to use this method, burning during the dry season is probably the easiest means of clearing.

COMPOSITAE

Psiadia punctulata (DC.) Vatke (=*Psiadia arabica* Jaub. & Spach)

DESCRIPTION

A small, more or less evergreen shrub, usually 1 —1.2 m high but occassionally reaching 1.8 —2.5 cm with green stems, simple, alternate leaves and heads of bright yellow flowers. The leaves are bright green, up to 12 cm long × 2.5 cm wide and tapered gradually at both ends, with a short stalk and a smooth margin. The flower heads are about 5 mm across and arranged in much branched, loose, more or less flat-topped inflorescences, up to 10 cm across, at the ends of the branches. Individual heads consist of a central group of tubular florets surrounded by a row of longer, strap-shaped florets and several rows of small, green bracts. The fruits are tipped with a group of whitish hairs and are wind-borne.

Distribution and importance

Psiadia is a native plant of wooded grassland, thickets and abandoned cultivation in many of the wetter districts of East Africa from the coast to about 1800 m. It is found in north-east Tanzania, in Zanzibar and in many parts of Kenya and Uganda. As a weed of grassland it is commonest in the Rift Valley district of Kenya, where it often grows in association with leleshwa (*Tarchonanthus*). The plant has a very bitter taste and is

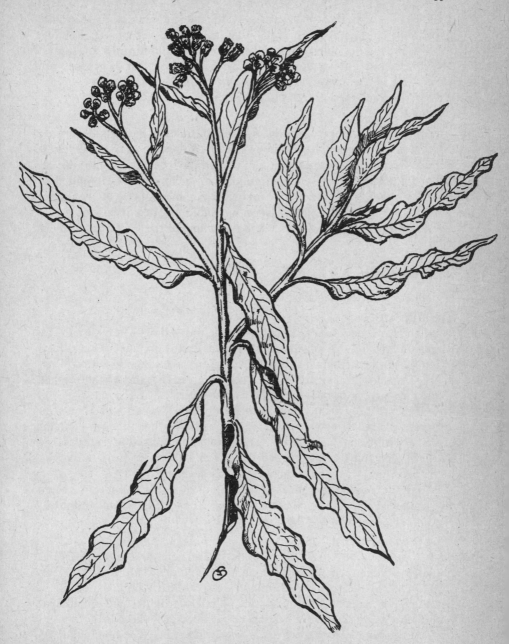

Figure 35. *Psiadia punctulata*

avoided by stock. Once it becomes established in grassland, gradually expanding clumps develop and under suitable conditions its copious seed production enables dense infestations to build up in a few years. Overgrazing or other factors which weaken the growth of grasses are favourable to the spread of *Psiadia,* but in the Naivasha area even apparently vigorous grassland is being extensively invaded.

Methods of control

Cutting gives temporary control, but regeneration is rapid and burning is relatively ineffective for the same reason. A number of trials with herbicides have been conducted in Kenya and have given encouraging results. Either by itself or in combination with picloram, 2.4-D applied as an overall spray will give a large measure of control of regrowth without injuring the grass. The information available at present suggests that the best time of application is at the beginning of the long rains in February.

COMPOSITAE

Tarchonanthus camphoratus L.   Leleshwa

DESCRIPTION

A much branched shrub or small tree reaching a height of 5 —6 m. The simple, alternately arranged, narrowly oval leaves up to 6 cm long smell of camphor when crushed and are covered on the underside with whitish hairs. The young twigs are also densely hairy so that the whole tree has a grey or silvery appearance. The inflorescences at the ends of the branches are brownish, much branched panicles, male and female on different plants. They consist of numerous flower heads, each containing 5 —12 tubular florets. Male florets are distinguished by their yellow anthers. The female have thread-like styles and are densely hairy, the hairs becoming wooly masses as the seeds develop.

When the main stems of leleshwa are cut or burnt, numerous shoots are produced from the swollen underground base of the trunk and thickets are formed in place of the original trees.

Distribution and importance

Leleshwa has a restricted distribution in Tanzania and Uganda (Karamoja), but forms a serious rangeland bush problem at altitudes between 1500 —2500 m in Kenya. It is commonest in the Rift Valley, especially in Baringo, Naivasha and Narok districts, and also troublesome in Laikipia, but does not appear to have become established east of the Rift.

Grass under leleshwa trees tends to be replaced by woody unpalatable plants. The trees are readily killed back to the ground by fire, but re-

N.S.

Figure 36. Tarchonanthus camphoratus

growth is very rapid and vigorous, and grass production is more serious-ly reduced in the thicket type of growth, sometimes to the point where the land becomes worthless for grazing. On the credit side, the roots make good charcoal and the foliage can be an emergency fodder reserve in times of exceptional drought.

Methods of control
Cutting only gives temporary control because of the rapid regrowth which occurs. Removal of the stumps is more permanently effective and, where charcoal can be made to cover the cost of clearing, is a practical treatment. After digging out, subsequent treatment is advisable to deal with regrowth from any remaining root fragments, otherwise the in-festation re-establishes itself. Chain-clearing followed by root-ploughing has also been successful on land suitable for wheat growing.

Chemical control of trees with a distinct main trunk has been achieved in the past with ester formulations of 2,4-D or 2,4,5-T (the latter no long-er marketed in East Africa) diluted with diesel oil and applied as a basal bark treatment. Such treatment, however, gives little success against thicket-type regeneration.

More recently, leleshwa has been found susceptible to herbicides con-taining picloram, and a formulation of picloram + 2,4-D applied as an overall foliar spray has been very effective in killing regrowth up to 2 m high. Treatments applied from May to July (in the Kedong Valley) gave the best results, with little or no further regrowth taking place. As growth slowed down later in the dry season and the short rains, the ef-fectiveness of spraying declined, and reached a minimum in January.

EBENACEAE

Euclea divinorum Hiern   Mukinyei
    (= *E. keniensis* R. E. Fries)

E. racemosa Murr. subsp. *schimperi* (A.D.C.) F. White
    (= *E. schimperi* (A.D.C.) Dandy)

DESCRIPTION

Evergreen shrubs or trees, *E. divinorum* sometimes attaining a height of 10 −12 m, *E. racemosa* rarely exceeding 3 m. Leaves simple and alter-nate, 5 −8 cm long by 2 −4 cm wide, shiny on the upper surface and somewhat leathery. The flowers are small, yellowish and arranged in racemes up to 4 cm long, the fruits globular, 5 mm across and green.

The species are best distinguished by the leaves which in *E. divino-rum* have reddish scales on the surface and are rhombic in shape, being widest near the middle. The leaves of *E. racemosa* lack the red scales and are widest near the apex. With both species not only is there copious regeneration from the base of the trunk after cutting, but suckers are also stimulated to grow from the roots up to a considerable distance from the original stem.

Distribution and importance

In their natural habitat *Euclea* species are constituents of evergreen forest or thicket in areas of relatively high rainfall at altitudes up to 2000 m, where they are of little economic importance. When the forest or thicket is cleared, however, they become much more prominent because of their apparently inexhaustible capacity for coppice and root sucker production. Even in bush where *Euclea* was originally only a minor constituent it often becomes the dominant species in the regrowth and, in addition to hindering the conversion of newly cleared bush into grazing land, also appears able to invade existing pasture.

Euclea occurs in suitable habitats throughout East Africa, but only in Kenya can it be classed as an important weed. Particularly serious infestations occur in Laikipia and the North Mara area. A low growing shrub form is also common in the coast hinterland.

Methods of control

Cutting *Euclea* regrowth has only a temporary effect and appears to encourage the proliferation of suckers which makes even destumping ineffective. Repeated cutting gradually reduces the size of the regenerating shoots, but it is doubtful whether even annual cutting will ever give a complete kill.

Chemical control is also difficult. Larger regenerating stems or stumps can be killed down to ground level with carefully applied basal-bark application of 2,4-D ester in diesel oil, and smaller regrowth (up to about 1 m high) can be killed back by overall sprays of the same chemical mixed with water. Such treatments delay the appearance of regrowth, but retreatment of shoots growing from the below-ground part of the stump and from roots is necessary at intervals to maintain the control. By using a formulation of picloram + 2,4-D as a foliage spray, the amount of retreatment needed can be reduced, but a treatment giving permanent control with a single application cannot yet be specified.

The effectiveness of chemicals is considerably influenced by the time of year at which they are applied and experiments in Kenya suggest that the best results with 2,4-D are obtained by spraying about 3 months after the main annual peak of rainfall.

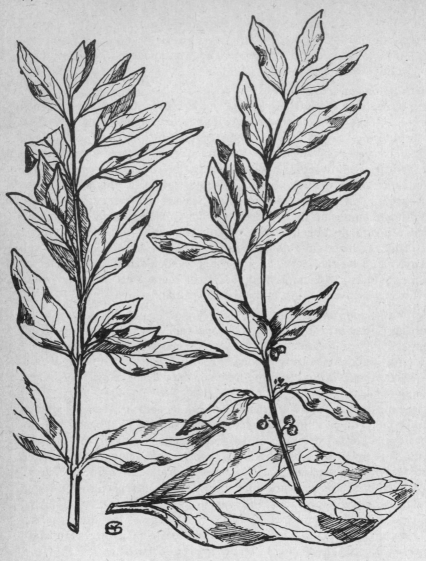

Figure 37 Euclea divinorum

MIMOSACEAE

Acacia brevispica Harms   Wait-a-bit Thorn
(= *A. pennata* auct.)

DESCRIPTION

A shrub consisting of numerous thorny, scrambling shoots arising from
a root crown at or just below ground level and reaching a height of up to
5 m. Unlike the majority of acacias, *A. brevispica* bears recurved thorns
scattered along the stem instead of in pairs at the base of the leaves. The
leaves are large and doubly pinnate, the leaflets small and very numerous. The small flowers are creamy white and grouped into round
heads about 1 cm across and clustered in panicles up to 30 cm long at the
ends of the branches. The fruits are flattened, oblong pods, up to 12 cm
long by 2.5 cm wide, brown when ripe and containing hard, brown seeds.

Distribution and importance
A. brevispica is widespread in Kenya, Tanzania and Uganda from the
coast up to about 2000 m, occurring in thicket and various types of bush
in the lower rainfall areas. In Kenya, it is reported as troublesome on
80,000 ha of grazing land in the Laikipia district, while large areas are
also affected in Machakos district. It is particularly common in the Karamoja district of Uganda.

Thickets containing this species are generally dense and difficult to
clear and, after clearing, the thorny shoots of *A. brevispica* are usually
the fastest growing constituents of the regeneration. It is commonly a
problem on abandoned cultivation and grazing land in thicket country
and, even when subjected to periodic burning, is able to regenerate repeatedly. It is also a common constituent of the regeneration where bush
(especially riverine thicket) has been cleared for tsetse control purposes.

Methods of control
A. brevispica survives repeated cutting, but small areas can be effectively cleared by the laborious process of digging out the root crowns.

For larger scale control chemicals show some promise, but it is almost
always necessary to clear the thicket before spraying is possible and
complete kills are difficult to achieve. The most effective chemical now
available is a combination of 2,4-D + picloram. Overall spraying may be
possible on young plants but for established thickets of *A. brevispica* it is
usually more practicable to cut first and apply the herbicide to the regrowth that develops.

The season of cutting and the height of the regrowth may be critical in
determining the effectiveness of treatment. The best results would be
expected from sprays applied when the initial growth of regenerating

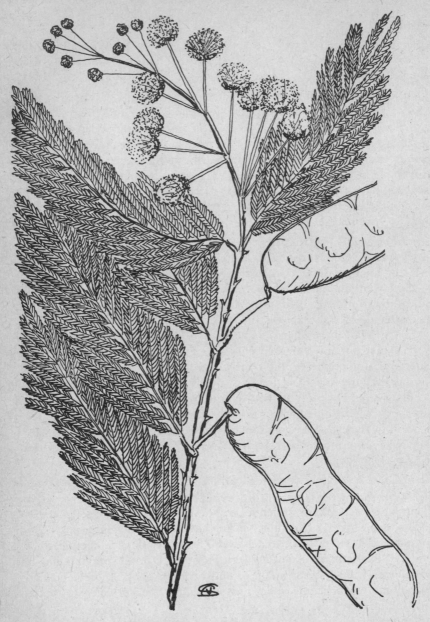

Figure 38. Acacia brevispica

shoots is slowing down. With *A. brevispica* this is normally 3 =6 months after cutting, depending on the season, by which time the shoots may be anything from 0.6 —2 m high. A single chemical treatment is unlikely to give complete control, but should greatly reduce further regeneration and a single follow-up treatment may be all that is required.

MIMOSACEAE

Acacia drepanolobium Sjöstedt	Whistling Thorn
A. gerrardii Benth.	
(= *A. hebecladoides* Harms)	
A. hockii De Wild.	White Thorn
(= *A. seyal* Del. var *multijuga* Bak. f.)	
A. lahai Benth.	Red Thorn
A. nilotica (L.) Del. subsp. *subalata* (Vatke) Brenan (= *A. subalata* Vatke)	
A. seyal Del.	
A. zanzibarica (S. Moore) Taub.	Coast Whistling-thorn

DESCRIPTION
Acacias are among the commonest of East African trees and many of the smaller species can occur as woody weeds of grassland. Those listed are the ones most commonly encountered in grazing land though over 50 additional species have been recorded in East Africa and many of these can also be troublesome on occasion.

The general appearance of *Acacias* needs no description. Features distinguishing the species mentioned are detailed in Table 1.

All the species mentioned have leaves made up of numerous small leaflets, with the exception of *A. zanzibarica* which has 3—10 pairs of large leaflets.

Distribution and importance
The *Acacias* listed are all common throughout East Africa, except for *A. zanzibarica* which is a coastal species and *A. lahai* which is restricted to altitudes above 1800 m and more commonly encountered in the highlands of Kenya than in Uganda or Tanzania. *A. drepanolobium* is the most troublesome species in Kenya, covering large areas of rangeland at medium altitudes. *A. hockii* is characteristic of areas with rather higher rainfall and is of particular importance in Uganda and western Kenya.

Acacias present a common problem in grassland. They are natural constituents of grassland communities and when the grazing intensity is increased by the introduction of domestic cattle, their numbers tend to

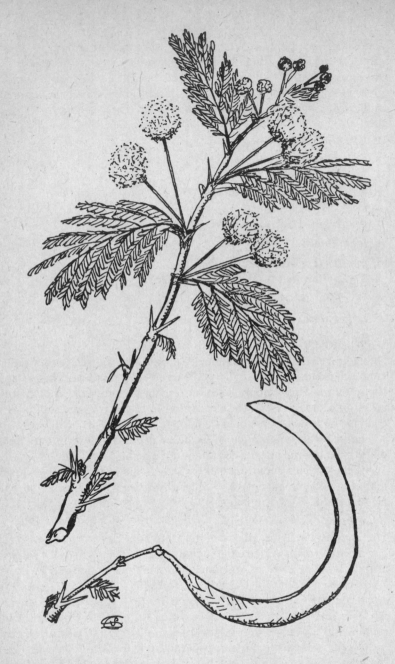

Figure 39. Acacia hockii

TABLE 1.

Distinguishing features of *Acacia* species

Name	Bark	Spines	*Flower* colour and shape of head	Pods
A. *drepanolo-bium*	grey to black, rough	long, straight, ant galls	white round	sickle-shaped to almost annular
A. *gerrardii*	grey to black, rough	short, straight	white round	sickle-shaped
A. *hockii*	red-brown to green-brown, peeling	short, straight	yellow round	sickle-shaped, narrow with wavy margin
A. *lahai*	grey to brown, rough	long, straight	cream elongated	± straight, broad, short
A. *nilotica*	brown to black, rough	long, straight, directed ± downward	yellow round	straight, oblong, densely hairy, fleshy
A. *seyal*	powdery white, yellow or red	long, straight	yellow round	sickle-shaped, narrow
A. *zanzibarica*	yellow to yellow-brown	long, straight, ant galls	yellow round	sickle-shaped, narrow

increase because of reduced competition, interruption of burning or other factors. The trees generally sprout rapidly from the base when felled and the thorny shoots near the ground interfere with grazing to a greater extent than the original trunks. Some species are still capable of producing regrowth, even when the stumps are dug out and *A. lahai* is especially troublesome because it can form root suckers up to a considerable distance from the stump when the tree is cut down.

Methods of control

Repeated cutting keeps *Acacia* regrowth under control but rarely gives a complete kill of the stumps. The possibility of chemical control depends partly on the species and partly on the stage at which the problem is tackled. The best results are obtaining if a treatment can be applied when the trees are first being cleared. The most effective of the chemicals currently available is picloram, sold in combination with 2,4-D. For application to freshly cut stumps small volumes of the undiluted concentrate have given good results (0.5 —1 ml on 8 —15 cm diameter stumps). The same material has also been effective injected through the bark every 8 cm round the base of standing trees.

On smaller trees and regrowth overall foliage sprays are more practicable than stump or basal treatment and good results have been achieved with sprays applied either at high volume rates or at low volume with a mistblower. Timing of foliar spraying is critical and with *A. drepanolobium* the best results have been obtained in the latter half of the long rains. Promising newer chemicals include triclopyr, applied as a foliage spray, and tebuthiuron, applied to the soil as pellets.

The results of chemical treatment have been variable with all *Acacias*. Success or otherwise depends to a large extent on the regenerative capacity of the tree, which is determined not only by the species but also by age, previous treatment, climatic and soil conditions. It is, therefore, difficult to classify species on the basis of degree of susceptibility or resistance. Among those which have been controlled with relatively little trouble are *A. nilotica, A. seyal* and *A. zanzibarica* (also the dry country *A. mellifera* and *A. reficiens*). On the other hand *A. hockii* and *A. lahai* are two types for which satisfactory control has not generally been obtained. Intermediate species, on which treatment has sometimes been most effective and sometimes quite ineffective include *A. drepanolobium* and *A. gerrardii.*

MIMOSACEAE

Dichrostachys cinerea (L.) Wight & Arn.   Chinese-lantern Tree
 (= *D. glomerata* (Forsk.) Chiov.)

DESCRIPTION
A much branched shrub or small tree up to 6 m high with the general appearance of an *Acacia,* but with single instead of paired spines and a two-coloured, cylindrical flower head, pink in the lower half (nearest the stem) and yellow in the upper. The spines are actually short branches and often bear leaves and flowers. The large, doubly pinnate, *Acacia*-like leaves are made up of numerous small leaflets about 2 mm wide. The pendulous flower heads are 5 —8 cm long and borne singly on 5 —8 cm stalks arising from the leaf axils. They consist of numerous small flowers about 3 mm long and develop into globular clusters of narrow,

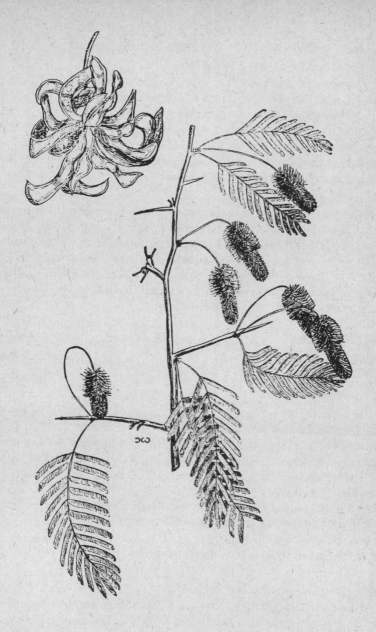

Figure 40. Dichrostachys cinerea

contorted, dark brown pods 5 —8 cm long × 1 —2 cm wide.

Distribution and importance

Dichrostachys occurs widely in Kenya, Tanzania and Uganda at altitudes up to 1500 m and also in the islands of Zanzibar and Pemba.

It is a suckering, thicket-forming species, which regenerates readily from the roots, and is particularly troublesome in abandoned cultivation and grassland in semi-arid areas. It is often an indicator of overgrazing and presents a very similar problem in grassland to that of some of the *Acacias* described in the previous section.

Methods of control

In its reaction to cutting and in its response to arboricides *Dichrostachys* is again similar to one of the more freely regenerating *Acacia* species. Where individual stems can be treated ester formulations of 2,4-D mixed with diesel oil and applied as basal-bark or stump treatments give good results. On denser, thicket-type growth foliar spraying with 2,4-D or 2,4-D/picloram in water can be effective and is best applied when the foliage is fully developed but not senescent.

VERBENACEAE

Lantana camara L.   Lantana, Tick-berry

L. trifolia L.   Sage Brush

DESCRIPTION

Lantana camara is a spreading, thicket-forming shrub reaching 2 —4 m high, with 4-angled, prickly stems. The aromatic leaves are arranged in opposite pairs and are about 7 cm long by 3.5 cm broad, with serrated margins and a rough upper surface. Individual flowers are small but are borne in flat, showy heads 3 —5 cm across. There are a number of colour forms, including several garden varieties, the most troublesome variety having pinkish purple flower heads with yellow centres. The fruits are small, black berries clustered into round heads.

L. trifolia is of similar habit but smaller, without prickles and with most of the leaves arranged in whorls of three. The flowers are pink and the fruits purple.

Distribution and importance

L. camara is widely distributed through the tropics and has been introduced into East Africa as a hedge plant. It is commonest on the coast and islands, but also appears to be spreading in the highlands of Kenya and certain other inland areas, and there are serious infestations in parts of Uganda.

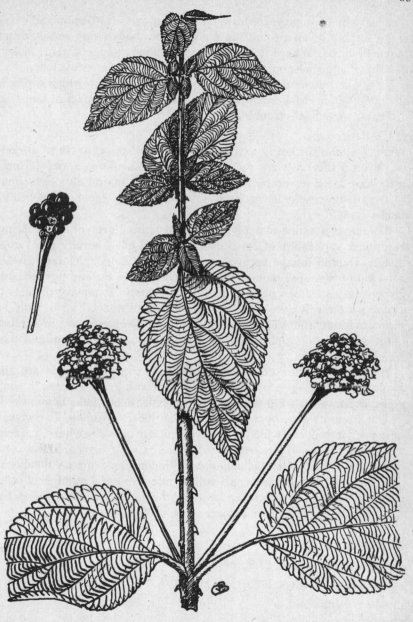

Figure 41. Lantana camara

Its principal habitats are waste land, abandoned cultivation and coastal thicket, but it can also invade pasture, the seeds being widely distributed by birds. Where it becomes established, impenetrable thickets develop giving shelter to tsetse. The leaves are poisonous to stock.

L. trifolia is a common constituent of secondary bush which grows in similar situations to *L. camara.* It is a less aggressive species, however, and causes much less trouble.

Methods of control

When *Lantana* thicket is cleared there is copious regrowth of suckers and even if the roots are dug out numerous seedlings appear. Herbicides have been tested in many parts of the world and although somewhat variable results have been obtained certain general conclusions can be made.

Where it is possible to apply a chemical without preliminary clearing, basal-bark application of 2,4-D ester in diesel oil generally gives good results. Overall foliage spraying with 2,4-D or 2,4-D/picloram gives a good kill of the tops, though some retreatment is usually needed, and glyphosate also shows promise as a foliage spray. Extensive thickets of *Lantana* are rarely accessible for spraying until opened up by cutting. Picloram formulations applied to stumps have been effective in Zambia, but it is often more practicable to allow the stumps to sprout and then spray the regrowth.

In addition to chemical control, biological control methods are also available for *L. camara* and have been used with great success in some areas. In Hawaii and Fiji particularly dramatic results have been achieved and attempts have been made to repeat these in Kenya. Two insects have been introduced which have given good results elsewhere, namely, a lace-bug (*Teleonemia scrupulosa*) which eats the leaves, and a seed-destroying fly (*Ophiomyia lantanae*). Under East African conditions these insects have not been particularly effective but a number of other insects are known to attack *Lantana* and the beetle *Uroplata girardi* has been effective in Zambia in association with *Teleonemia.*

VERBENACEAE

Lippia javanica (Burm. f.) Spreng   Sage Brush

(= *L. asperifolia* A. Rich.)

L. ukambensis Vatke

DESCRIPTION

The *Lippia* species, of which several very similar types occur in East Africa, are closely related to *Lantana,* but can be distinguished by their dry instead of fleshy fruits and their rounded heads of densely packed, white flowers. Both the above species are aromatic shrubs growing to a height of 1 —2 m with opposite, simple, finely-toothed leaves, which are rough above, densely hairy below and prominently nerved. The small, flowers form heads about 15 mm long which are borne on long stalks arising in the leaf axils.

The two species can be distinguished by the bracts round the base of the flower heads. In *L. javanica* they are relatively broad, shortly pointed and densely hairy, whereas in *L. ukambensis* they are narrower, have a long tapering point and are only thinly hairy. When the flowers are open, *L. javanica* heads are generally somewhat less than 12 mm long while *L. ukambensis* heads are generally slightly more.

Distribution and importance
Both species are indigenous and distributed quite widely through East Africa as constituents of secondary bush, particularly in the 1200—2000 m altitude range. In parts of northern Tanzania they form dense thickets in mountain grazing land. They are also troublesome in grazing areas north of Kilimanjaro and in the Rift Valley. Like *Lantana* they become re-established rapidly after cutting by means of sprouting and the germination of seed.

Methods of control
A number of herbicide trials have been conducted on *L. javanica* in East Africa and, when applied as overall sprays, ester formulations of 2,4-D have given good results. The time of year at which the application is made is important and in northern Tanzania the best results—almost complete kill—were obtained in May, during the long rains. August treatment was much less effective.

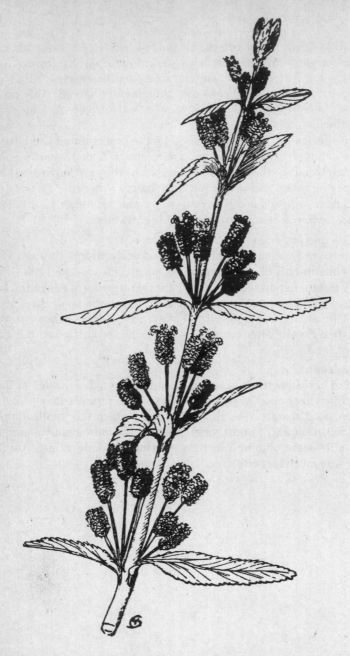

Figure 42. Lippia javanica

HERBACEOUS WEEDS

AMARANTHACEAE

Achyranthes aspera L. Devil's Horsewhip

DESCRIPTION

A coarse herb, sometimes growing in arable land as an annual, but under suitable conditions lasting for a number of years and reaching a height of 1.2 —1.5 m. Stems tough, becoming woody at the base. Leaves opposite, simple and ovate, up to 10 cm long by 8 cm wide, tapering to a point at both ends and shortly stalked. Individual flowers are small, pink or greenish in colour and form narrow, elongated terminal spikes up to 60 cm long. As the flowers age, they bend downwards and become pressed closely against the stem. The bracts surrounding the flowers in the fruiting stage have sharp, pointed tips making the heads spiny to the touch. They also cause the fruits to stick to the hair of animals, clothing etc. and so assist dispersal.

Distribution and importance

An indigenous plant found in many parts of Africa and widely distributed through East Africa in hedges, thickets and shaded habitats generally. It occasionally occurs as an arable weed but more often as a weed of waste ground. In Kenya it is recorded as of some importance in the Machakos and Kitale districts; in Tanzania it is occasionally troublesome in the Ilonga district and the Southern Highlands.

Methods of control

Achyranthes is moderately resistant to 2,4-D and MCPA. In the young seedling stage a reasonable kill can be obtained with rates of the order of 1 kg per ha, but resistance increases rapidly with age and older plants require 2 kg per ha or more.

AMARANTHACEAE

Alternanthera pungens H.B.K. Khaki Weed
  (= *A. repens* (L.) O. Ktze.)

DESCRIPTION

A low-growing perennial with prostrate branches spreading from a thick tap root to a length of 30 —60 cm and rooting at the nodes. The stems are hairy and often reddish in colour. The leaves are arranged in opposite

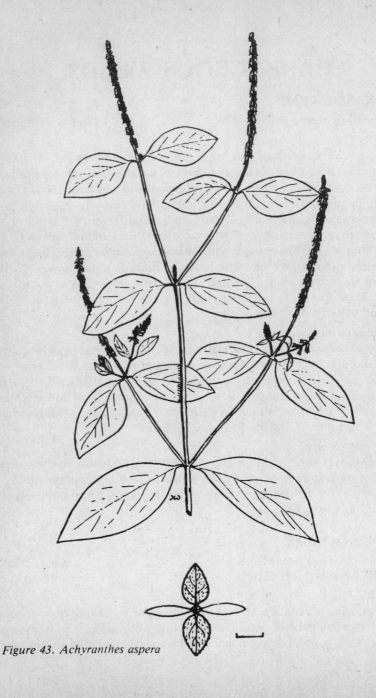

Figure 43. *Achyranthes aspera*

Figure 44. Alternthera pungens

pairs of unequal size, the larger ones up to 2 cm broad by 4 cm long. They have short petioles, obovate blades tipped with short points and entire margins. The inconspicuous, straw-coloured flowers are clustered in dense, sessile heads in the axils of the leaves. Individual flowers are compressed and their structure difficult to make out, but the five unequal-sized perianth segments and the three bracts associated with each flower are all tipped with sharp spines. The fruits adhere to the skin of men and animals, to the soles of shoes and to motor tyres and are readily distributed.

Distribution and importance
A plant probably introduced to Africa from South America, but now found throughout the tropics, *Alternanthera* is common throughout East Africa on a wide range of soil types. It is not uncommon as a weed of arable land, but is more characteristic of roadsides, pathways and waste land and often occurs in lawns where its spiny fruits can be troublesome in the feet of children and dogs.

Methods of control
Alternanthera is slightly susceptible to 2,4-D but can only be controlled in the seedling stage using relatively high rates. Formulations containing picloram are more effective and can be used in uncropped land, as can amitrole. In arable crops there is little alternative to cultivation at present though cereal herbicides containing clopyralid offer a chance of selective chemical control.

AMARANTHACEAE

Amaranthus dubius Thell.
A. graecizans L. (= *A. angustifolius* Lam.)
A. hybridus L. (= *A. hypochondriacus* L.)   Pigweed
A. lividus L. (= *A. blitum* L.)
A. spinosus L.   Spiny Pigweed
A. thunbergii Moq.

DESCRIPTION
A group of annual herbs growing to a height of 0.6 —1 m and all similar in general appearance. The leaves are arranged alternately and are simple, with entire margins, the blades mostly narrowing gradually into the distinct stalk. There is usually a minute point at the apex. Individual flowers are greenish and inconspicuous and are clustered into dense heads, sometimes elongated and cylindrical. The inflorescence may be interrupted and leafy to the tip or the upper part may be leafless and form a terminal, spike-like panicle. Male and female flowers are sepa-

rated but borne on the same plant. Fruits each contain a single black seed.

Several species other than those listed above may be important locally. The species mentioned, however, include the ones most commonly found in East Africa. They may be distinguished by the characters given in the·following table (though careful dissection and inspection of the flower heads are needed to make out details of the flower structure).

Distribution and importance

Most of the species listed have a wide distribution in tropical countries and appear to be generally distributed through East Africa over a wide range of altitude. They are typically weeds of arable land, producing seed abundantly and often occurring in very large numbers. *A. thunbergii* is very common in Zimbabwe, but is less common than the other

TABLE 2.

Distinguishing features of *Amaranthus* species

Species	Inflorescence	Perianth segments of female flower (P)	Fruit	Other characters
A. dubius	Leafy to tip	5 tip blunt	Shorter than P, splitting round middle	Leaf base broadly triangular
A. graecizans	Leafy to tip	3 tip very shortly pointed	Longer than P, splitting round middle	—
A. hybridus	Terminal part leafless	5 tip acute	Longer than P, splitting round middle	—
A. lividus	Terminal part leafless	3	1½-2 times length of P, smooth not splitting	Plant glabrous, leaf apex bilobed
A. spinosus	Terminal part leafless	5 tip spiny	—	Paired spines at base of leaves
A. thunbergii	Leafy to tip	3 tip with long slender point	—	—

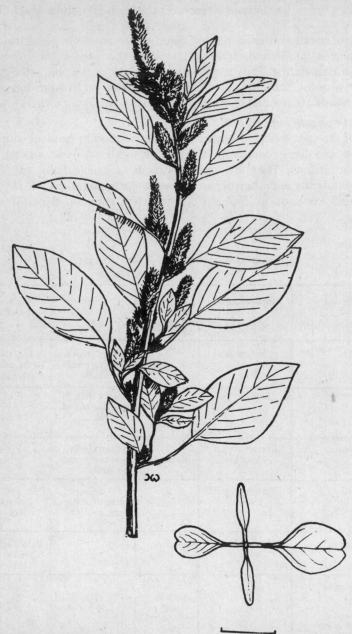

Figure 45. Amaranthus hybridus

species in East Africa, having been recorded only from a few localities in Kenya. *A. lividus* is not of great importance as a weed, but is commonly cultivated as a form of spinach in Kenya and Tanzania.

Methods of control
Seedlings of *Amaranthus* are sensitive to 2,4-D, MCPA and other herbicides of the growth-regulator type at the doses normally used in cereal crops. 2,4-D can also give good control as a pre-emergence treatment. In addition they are susceptible to pre-emergence application of many soil applied chemicals including the triazines, substituted ureas, amides, dinitroanilines and EPTC.

AMARANTHACEAE

Cyathula polycephala Bak.
C. cylindrica Moq. (= *schimperana* Moq.)
C. orthacantha (Asch.) Schinz
C. prostrata (L.) Blume
C. uncinulata (Schrad.) Schinz (= *C. globulifera* Moq.)

DESCRIPTION

A group of scrambling or erect herbs with opposite, stalked, simple leaves and inconspicuous flowers clustered in dense, prickly, whitish or straw-coloured heads which form burrs in the fruiting stage. Each head is made up of several fertile flowers and a large number of sterile ones, with which are associated rigid spines which may be either straight or hooked.

    C. polycephala is a scrambling perennial reaching a height of about 1 m. The leaves are hairy, up to 8 cm long × 4 cm broad with a rhomboid blade and pointed tip. The inflorescence is made up of globular clusters, 1 cm across and arranged in an interrupted, unbranched spike up to 20 cm in length. The spines of the sterile flowers are hooked at the tip and whitish.

    C. cylindrica has been confused with *C. polycephala* but differs in its much shorter, densely cylindrical and uninterrupted inflorescence, about 1 cm across. The stem is sparsely clothed with long, downward-directed hairs and this species has been reported to arise from a large potato-like tuber.

    C. prostrata has a slender, cylindrical flower spike, 5 mm across, up to about 10 cm long and interrupted towards the base. Stem and leaves are densely hairy and the spines yellowish.

    In *C. uncinulata* the flower heads are borne singly rather than being arranged in a spike and are globular, about 2 cm in diameter.

    C. orthacantha differs from the other species in being an erect annual

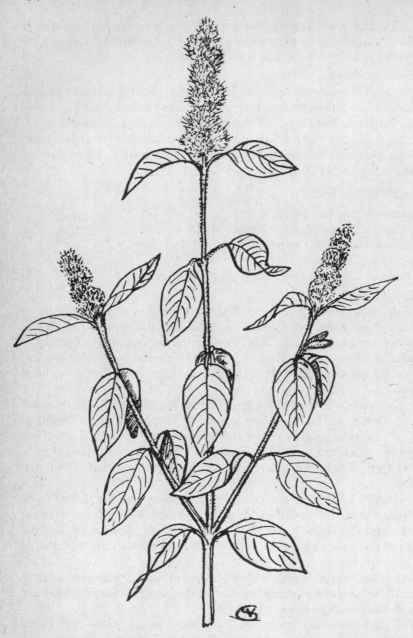

Figure 46. *Cyathula cylindrica*

and in the straight, unhooked, yellow spines of the sterile flowers. The inflorescence is short (about 5 cm long), dense and broad.

Distribution and importance

The commonest species is *C. polycephala* which occurs, mainly in grassland, at altitudes of 1500 —2500m in Kenya, Tanzania and Uganda. It is reported as one of the most troublesome weeds in Kikuyu grass pastures in Kenya, where its burrs cause considerable damage to the wool of sheep.

    C. cylindrica and *C. uncinulata* have a similar distribution in high altitude grassland and forest margins, but are less common than the above, while *C. orthacantha* is widespread in grassland, at the 800 —1400 m altitude range. *C. prostrata* is a forest herb in many parts of the tropics besides East Africa and is recorded as a weed of cultivation in forest areas of Uganda.

Methods of control

The perennial *Cyathula* species regenerate vigorously when cut. Information on susceptibility to chemicals is scanty but 2,4-D and MCPA appear to be relatively ineffective on *C. cylindrica*. It is likely that formulations containing dicamba, picloram or clopyralid would give more lasting control. Application of glyphosate by means of a 'wiper-bar' applicator would also be worth trying.

BORAGINACEAE

Cynoglossum coeruleum DC.   Forget-me-not
C. geometricum Bak. & Wright
C. lanceolatum Forsk.

DESCRIPTION

C. coeruleum is an erect, much branched, roughly hairy annual, growing to about 60 cm with alternate, simple leaves and bright blue flowers. The stem leaves are up to 6 cm long × 1 cm wide, gradually narrowed to the pointed apex and the sessile base; the lower leaves are larger and stalked. The flowers are 5 mm across and are borne on 5 mm stalks along the branches of the loose, terminal inflorescence, each branch bearing up to 10 flowers. As it ages the inflorescence elongates and the flower stalks bend downwards. Each flower consists of a 5-toothed calyx, a corolla of 5 united, rounded petals, 5 stamens and an ovary developing into a fruit made up of 4 nutlets. The mature fruits are 5 mm across and covered with short, hooked bristles.

    In *C. geometricum* the lower leaves are not distinctly stalked. The flowers are pale blue, rather less than 5 mm across, and form a looser

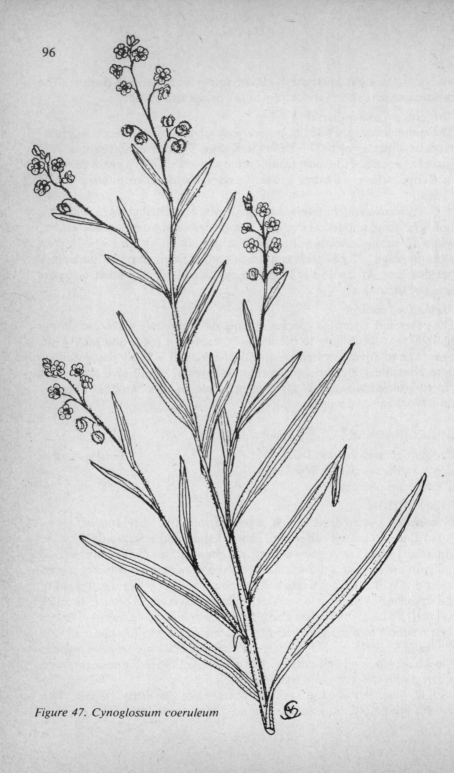

Figure 47. Cynoglossum coeruleum

inflorescence with up to 15 flowers per branch. The flower stalks are about 3 mm long and the fruit, although about the same size as in *C. coeruleum*, bears hooked bristles only round the edges of the nutlets and along the central line.

C. lanceolatum grows to a somewhat greater size than the other two species and has broader leaves. The flowers are usually white (though sometimes pale blue), 3 mm across and have very short stalks. The inflorescence branches bear up to 20 flowers, more widely separated than in *C. coeruleum*, while the fruits are 3 mm across and uniformly covered with bristles.

Distribution and importance

The three species are indigenous in East Africa and occur widely at altitudes from 1200 —2000 m. *C. coeruleum* and *C. lanceolatum* are found in all three countries, while *C. geometricum* is commonest in Kenya and Uganda. All the species are common on waste land and roadsides and are often encountered as arable weeds, especially in the Naivasha and Sotik districts of Kenya. They also occur in pasture land where the burrs can be troublesome in wool.

Methods of control

Cynoglossum species are relatively susceptible to 2,4-D and MCPA and can be effectively controlled as young seedlings in cereal crops or grassland. 2,4-D also gives control as a pre-emergence treatment. There is little information on the effects of the newer growth-regulator type chemicals or of soil-applied herbicides.

BORAGINACEAE

Heliotropium steudneri Vatke (= *H. eduardii* Martelli)

H. undulatifolium Turrill

H. ovalifolium Forsk.

H. indicum L.

A group of hairy herbs growing to a height of about 1 m with alternate, simple leaves and small white flowers arranged in long, one-sided spikes, which are curled backwards at the tip.

H. steudneri and *H. undulatifolium* are perennials with extensive underground rhizome systems. *H. steudneri* has leaves 3 —6 cm long, shortly stalked and ovate in shape, with a pointed tip and flat margin. The flower spikes are borne in the axils of the upper leaves and are 6 —12 cm long. The flowers are numerous and arranged in 2 closely packed rows. They consist of 5 joined sepals, a corolla of 5 joined petals with 5 stamens attached to its tube and a 4-celled ovary. *H. undulatifolium* differs mainly in its narrower leaves 2 —4 cm long, with the margins wavy in the plane at right angles to the surface and in its shorter spikes up to 4 cm long.

The other two species are annuals similar in general appearance to the perennials. *H. ovalifolium* is greyish in colour, with ovate leaves up to 4 cm long, which are blunt at the tip, narrowed gradually to the short stalk and have a smooth margin. *H. indicum* is larger and has broadly ovate leaves 8 —12 cm long, narrowed sharply into the stalk and with an irregularly toothed margin. The spike sometimes reaches a length of 30 cm.

Distribution and importance

The perennial species are the most important as weeds and, of the two, *H. steudneri* is probably the more often encountered. It occurs mainly in grassland at altitudes of 1200 —1800 cm in all three countries, though less commonly in Uganda. It is often found on roadsides and waste ground and sometimes becomes established in arable land, where it is a bad weed. It is particularly troublesome on volcanic soils near Arusha in Tanzania. *H. undulatifolium* occupies similar habitats over a wider range of altitudes, from 500 m in the Usambaras to 2700 m on the Kinangop, but is not recorded from Uganda.

The annual species present a less serious weed problem. They occur most frequently near rivers and lakes, but occasionally spread into cultivated areas. *H. ovalifolium* is found both in Tanzania and Kenya, *H. indicum* mostly in Tanzania.

Methods of control

What little information is available suggests that the two annual species are moderately susceptible to 2,4-D and MCPA, and can be effectively controlled in the seedling stage. The perennial species are more difficult. With *H. steudneri* the top growth can be killed back quite easily, but there is generally only a poor kill of the underground rhizomes and regrowth is rapid. *H. undulatifolium* probably reacts in much the same way. It is probable that non-selective control could be obtained with such chemicals as amitrole or glyphosate and treatment with hormone formulations containing dicamba, picloram or clopyralid would be worth trying in grassland.

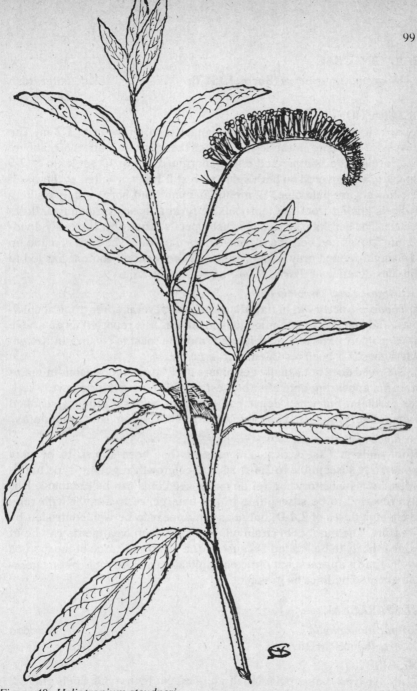

Figure 48. Heliotropium steudneri

BORAGINACEAE

Trichodesma zeylanicum (Burm. f.) R. Br.   Kuenstler Bush
      Late Weed

DESCRIPTION

A coarsely hairy, erect, branched, annual herb, growing to 1.2 m. The leaves are opposite in the lower part of the plant and alternate above. They are sessile, simple and entire-margined, up to 10 cm long × 2.5 cm wide, and covered on both sides with stiff hairs, swollen at the base. The flowers are pale blue, 12 mm in diameter and borne singly on long pedicels arising from the leaf axils to form a loose, spreading inflorescence. The corolla consists of 5 fused petals. The calyx is deeply divided into 5 lobes and becomes much enlarged in fruit. The fruit is made up of 4 single-seeded grey nutlets, whose 30 per cent oil content has led to the consideration of *Trichodesma* as an oil seed crop.

Distribution and importance

Trichodesma occurs in India, Mauritius and several other tropical countries, and appears to be indigenous in Africa. It is reported as an arable weed in many parts of Tanzania, and from the coast to Nyanza in Kenya. It has not so far been recorded from Uganda.

The weed does not usually germinate until late in the season, in maize often not appearing until after harvest. At this period it is not a very serious problem, but when dense can interfere with subsequent seedbed preparation. It has also been noted in crops of sorghum and groundnuts.

Methods of control

From trials in East Africa, *Trichodesma* has been shown to be only moderately susceptible to 2,4-D and other growth-regulator type herbicides, but satisfactory control of young seedlings can be obtained. It is also reported to be susceptible to pre-emergence treatment with relatively high doses of 2,4-D, but does not appear to be well controlled by simazine. The number of situations in which such treatments can be of value is likely to be limited because of the advanced stage of most crops by the time it appears, but directed application of 2,4-D or a contact herbicide may sometimes be possible.

CAPPARACEAE

Cleome monophylla L.   Spindlepod
C. hirta (Klotzsch) Oliv.

DESCRIPTION

C. monophylla is an erect annual, up to 60 cm high, with finely ridged, hairy stems, alternate leaves and pink flowers. The leaves are simple

Figure 49. Trichodesma zeylanicum

Figure 50. Cleome monophylla

and narrowly oblong, up to 8 cm long with a 1 cm stalk. The flowers are about 1 cm across and borne singly on stalks arising in the axils of the upper leaves to form few-flowered, terminal racemes. They are made up of 4 separate sepals, 4 petals and 4—6 stamens arising from below a simple ovary. The ovary elongates greatly in fruit to form a narrow, many-seeded capsule up to 10 cm long and opening from below by two valves.

Several other *Cleome* species occur in East Africa and may occasionally be of importance as weeds. Mostly they have digitately compound leaves. The commonest is *C. hirta,* a species covered with sticky, glandular hairs and with 5—9 narrowly lanceolate leaflets at the end of a long stalk. The flowers are similar to those of *C. monophylla* but there are 10—12 stamens.

Distribution and importance

A common arable weed in tropical Africa, *C. monophylla* is of widespread occurrence in the East African countries from sea level to about 1500 m. *C. hirta* is distributed as widely, but is less often reported as a weed. It has been noted in maize and cotton in the Kilosa district of Tanzania and was common in the groundnut areas at Kongwa.

Methods of control

These two *Cleome* species are readily controlled by pre-emergence treatment with 2,4-D or MCPA at normal rates and young seedlings can also be killed by post-emergence sprays. Control in maize appears to be quite straightforward, but, in groundnuts, MCPB and 2,4-DB have given poor control of *C. monophylla*.

CAPPARACEAE

Gynandropsis gynandra (L.) Briq.   Spider Flower
  (= *G. pentaphylla* DC.)

DESCRIPTION

A much branched, erect, annual herb growing to a height of 0.6—1 m with showy white flowers and numerous sticky, glandular hairs. The leaves are alternate, long-stalked and palmately compound, and bear 5 (rarely 3—4) sessile, obovate leaflets up to 5—8 cm long with entire or finely toothed margins. The flowers are about 2 cm across, arranged in long, terminal racemes, and are borne on 1 cm stalks which arise singly in the axils of small, leafy bracts. There are 4 sepals, 4 narrow clawed petals and 6 stamens with long, purple filaments arising from a much elongated receptacle. The ovary also becomes greatly elongated in fruit and eventually forms a spindle-shaped capsule 8—10 cm long.

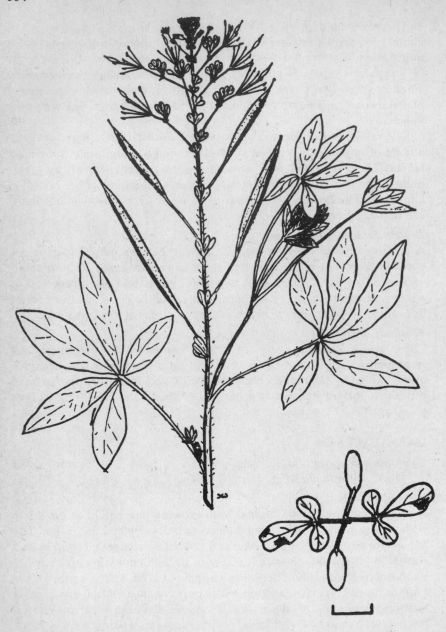

Figure 51. Gynandropsis gynandra

Distribution and importance
Gynandropsis is found in many tropical countries and is common in Kenya, Tanzania and Uganda as well as in Zanzibar and Pemba. In Kenya it reaches an altitude of at least 1800 m. Although common in arable land it rarely appears to grow in large enough numbers to be a major pest, but has been reported as troublesome in the Machakos district of Kenya. The leaves are edible and are often gathered as a green vegetable.

Methods of control
Good control of *Gynandropsis* can be obtained with 2,4-D, MCPA or other growth-regulator type herbicides, applied either as pre- or post-emergence treatments. It is also susceptible to pre-emergence application of low doses of simazine and many other residual, soil-applied chemicals.

CARYOPHYLLACEAE

Corrigiola littoralis L. subsp. *africana* Turrill   Strapwort

DESCRIPTION
A much branched, perennial herb with prostrate branches up to 50 cm long. Leaves spirally arranged, up to 2 cm long × 3 mm wide, acute at the apex and gradually narrowed to the base. At the point of junction with the stem are a pair of small, white, membranous, pointed stipules. The inconspicuous flowers are clustered into small heads near the ends of the branches. They are borne on very short stalks and consist of 5 rounded sepals, 5 white petals shorter than the sepals and alternating with 5 stamens, and a one-seeded ovary with 3 sessile stigmas.

Distribution and importance
The African subspecies of this weed occurs typically as a perennial weed of arable land, between 1200 and 3000 m. In Kenya it is common in various parts of the highlands, including the Uasin Gishu, Laikipia and Molo districts. It has been recorded in the northern part of Tanzania and in the Usambaras, but does not appear to occur in Uganda.

It has been noted particularly as a weed in cereals and grass leys and sometimes occurs in pyrethrum.

Methods of control
Information on the reaction of this weed to herbicides is scanty. It does not appear to be very sensitive to applications of 2,4-D or MCPA but formulations containing dicamba are reported to be effective.

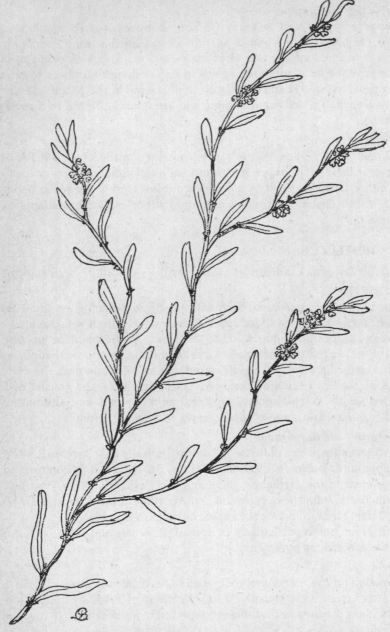

Figure 52. Corrigiola littoralis

CARYOPHYLLACEAE

Silene gallica L   Campion
   (= *S. anglica* L.)

DESCRIPTION

An erect, annual herb growing to a height of 50 cm and covered with spreading hairs. Leaves in opposite pairs, simple, oblanceolate, gradually narrowed to the base and up to 8 cm long × 2.5 cm broad, with a rounded apex bearing a small, pointed tip. The white or pinkish flowers are borne singly on short stalks in the axils of one of each of the upper pairs of leaves, to form a one-sided, few-flowered, loose inflorescence. The flowers are about 1 cm long; the calyx tubular, with 5 teeth and inflated in fruit. The 5 petals have a long claw and an expanded terminal lobe extending a short distance beyond the calyx tube. There are 10 stamens and the ovary generally has 3 styles.

Distribution and importance

Silene gallica originates from Europe, but has now been introduced, presumably with cereal seed, to many parts of the world. In East Africa it is a highland weed and is most troublesome in the wheat growing districts of Kenya, especially in the Uasin Gishu area. It has also been recorded at Kigezi in the west of Uganda and at Lushoto in Tanzania. Wheat is the crop most commonly infested, but is also occurs in maize, pyrethrum and other arable crops.

Methods of control

Like other *Silene* species, *S. gallica* is moderately resistant to herbicides of the growth-regulator type. Doses of 2,4-D and MCPA employed in cereal crops will give a severe check, if applied at a very early stage, before the seedlings exceed a height of 4 cm, but older plants are little affected. Mecoprop, either alone or mixed with other cereal herbicides, is used in wheat and barley. Of the soil-applied chemicals atrazine is effective pre-emergence and trifluralin can be used in a number of legume and other crops as a pre-sowing, incorporated treatment.

CARYOPHYLLACEAE

Spergula arvensis L.   Spurrey

DESCRIPTION

A slender, annual herb, branched at the base and usually 15 –30 cm high, but sometimes reaching 60 cm. The linear leaves are up to 6 cm long, grooved beneath and arranged in opposite pairs, but are apparently whorled because of the presence of axillary leaf tufts. The white flowers are 5 mm across on stalks 1 –2 cm long and form a loose, terminal

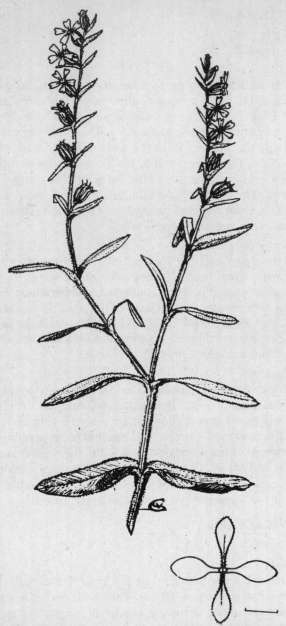

Figure 53. Silene gallica

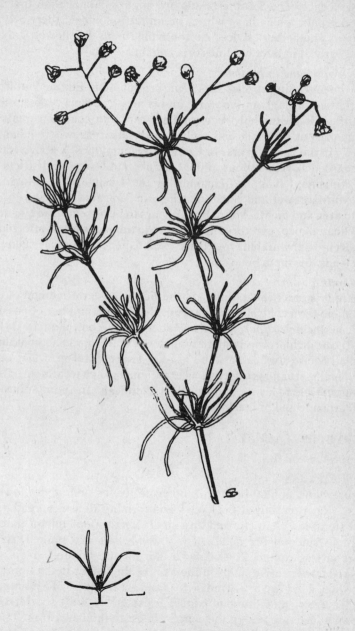

Figure 54. Spergula arvensis

inflorescence. The 5 entire petals are slightly longer than the 5 blunt-ended sepals, which have white, membranous edges: there are 5 or 10 stamens and 5 short styles. As the fruit ripens the flower stalk bends downwards, but later becomes erect again.

Distribution and importance
Like the preceding species, *Spergula* originated in Europe, but has been distributed throughout the world and is now a serious problem in many countries. It presumably reached East Africa in seed of cereals and is commonest on acid soils in the wheat growing areas, where it has spread rapidly in the last 15 years. In Kenya it is found mostly above 1500 m. It is regarded as very serious in the Laikipia and Naivasha districts and on the Kinangop. It is also common in the Uasin Gishu, Molo, Embu, Elgeyo-Marakwet and Nandi districts. In Tanzania it occurs in the Lushoto area and on Mt. Meru, but appears to be absent from Uganda.

Wheat, maize, pyrethrum and peas are the crops worst affected, but it can occur in any arable crop. Once established, its copious seed production leads to a rapid build-up.

Methods of control
Where it is practicable, liming is the best long-term method of control. For short-term control in arable crops there are a number of possibilities with herbicides. *Spergula* is not adequately controlled by MCPA or 2,4-D but dichlorprop is more effective, as are products containing dicamba. Metoxuron can also be used to give control in wheat or barley and pendimethalin is effective as a pre-emergence treatment. The weed is susceptible to a variety of residual chemicals, including diuron, linuron, atrazine and propachlor.

CARYOPHYLLACEAE
Stellaria media (L.) Vill.   Chickweed

DESCRIPTION
A sprawling, annual herb with opposite leaves and white flowers, the weak stems rooting at the nodes and forming dense, spreading mats. The stems are characterized by a single longitudinal line of hairs which changes sides at the nodes. The lower leaves are stalked, the upper more or less sessile. The blades are ovate and pointed at the tip. The stalked flowers arise singly in the axils of the leaves in the upper part of the stem. They are 6 —10 mm across and are made up of 5 pointed sepals with narrow, membranous edges, 5 petals, slightly shorter than the sepals and divided down the centre almost to the base, from 3 to 10 stamens and 3 styles arising from a capsule opening into 6 parts.

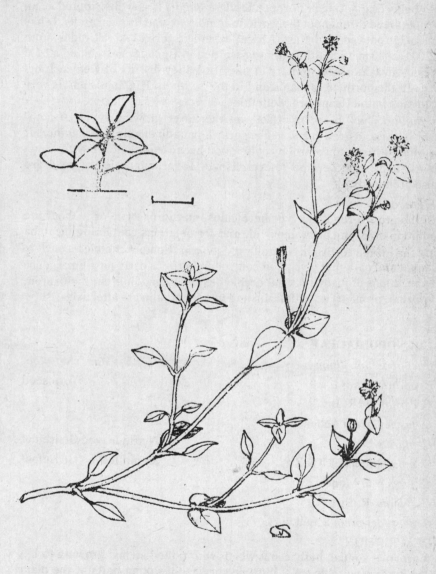

Figure 55. Stellaria media

Distribution and importance

Probably originating in Europe, *Stellaria media* is now distributed as an arable weed throughout the world in temperate and tropical areas. It has been introduced into East Africa, where it thrives at altitudes from 1200 —2000 m and, although not recorded in Uganda, has gained a hold near Naivasha and elsewhere in the Rift Valley district of Kenya. It occurs in the north of Tanzania and in the Southern Highlands and is very common in the Usambara Mountains.

Stellaria has its worst effects on the lower growing types of crop, which its rapid growth and dense sprawling habit enable it to smother. It is often difficult to control by cultivation because of the ease with which portions of stem become re-established. Large quantities of seed are also produced.

Methods of control

Of the growth-regulator type herbicides, mecoprop, with or without the addition of ioxynil or bromoxynil, and formulations containing dicamba are the most effective in controlling *Stellaria*. Contact chemicals such as diquat and paraquat give good control and pre- or early post-emergence applications of many residual herbicides, including linuron, metoxuron, atrazine, prometryne, trifluralin and pendimethalin are affective.

CHENOPODIACEAE

Chenopodium album L.	Fat Hen, Goosefoot
C. ambrosioides L.	Wormseed
C. carinatum R. Br.	
C. fasciculosum Aellen	
C. murale L.	Nettle-leaved Goosefoot
C. opulifolium Koch & Ziz	Round-leaved Goosefoot
C. procerum Moq.	
C. pumilio R. Br.	
C. schraderanum Schult.	

DESCRIPTION

A group of annual, herbaceous weeds with ribbed stems, growing to 1 — 2 m high on suitable soils. In some species the young parts of the plant are covered with grey or whitish, mealy hairs; others have yellow glandular hairs and an aromatic smell when crushed. The leaves are alternate and stalked, the flowers small and greenish, usually crowded into

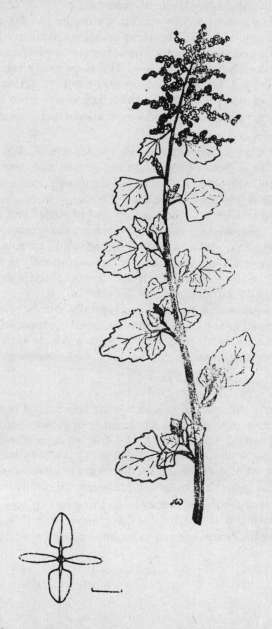

Figure 56. Chenopodium opulifolium

dense rounded clusters variously arranged.

There is a great deal of variation within the species and they are not always easy to identify. The most constant characters are those of the size, shape and surface markings of the seeds which can only be determined under high magnification. The more easily seen characters are listed in Table 3. They may not always permit certain identification, however, and in cases of doubt the fuller key given in the Flora of Tropical East Africa, volume Chenopodiaceae, should be consulted.

Distribution and importance

The two most widely distributed species are *C. opulifolium* and *C. schraderanum,* both of which are indigenous and common at altitudes from about 900 —2000 m. *C. murale* is spread over a large part of the world, but is less common in East Africa. *C. fasciculosum* and *C. procerum* are restricted to the eastern part of Africa and are locally common. *C. ambrosioides* occurs throughout the tropics, but appears to have been introduced into Kenya, Uganda and Tanzania. The other three species are much less common, being found mostly in the Kenya highlands, *C. album* originating from Europe and *C. carinatum* and *C. pumilio* from Australia.

The *Chenopodium* species are typically weeds of arable land and waste ground near human habitation. Seed is produced in large quantities and dense infestations are common in a wide variety of crops. The strongly aromatic species can be very objectionable in hay and fodder crops.

Methods of control

C. album is the species on which most information is available. Young plants are susceptible to the whole range of growth-regulator type herbicides, including 2,4-D, MCPA, 2,4-DB, mecoprop and dicamba. They are also susceptible to contact chemicals such as ioxynil, diquat and paraquat, while germinating seedlings are readily controlled by residual chemicals of the triazine, substituted urea and dinitroaniline groups.

There is much less information on the other species. Most appear to be susceptible to the herbicides listed above, but *C. schraderanum* and *C. ambrosioides* are reported to require rather higher doses of 2,4-D and MCPA.

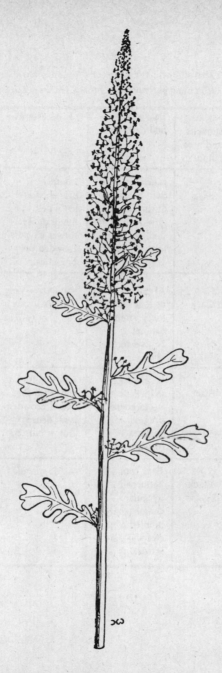

Figure 57. Chenopodium schraderanum

116

TABLE 3.

Distinguishing features of *Chenopodium* species

A. Young parts of plant mealy, leaves not aromatic.

Species	No. of stamens and seed colour	Leaf shape and margin	Inflorescence	Other characters
C. album Kenya, northern Tanzania. Not common.	5 Black	Much longer than broad. No distinct basal lobes. Up to 10 shallow, blunt teeth.	Loosely branched and leafy below. Small dense flower clusters crowded on branches.	Often red tinged. Branches often erect.
C. Opulifolium Kenya, Uganda, Tanzania. The commonest.	5 Black	About as broad as long. Basal lobes often distinct. Several teeth above.	As above. Very densely mealy.	Branches usually diverging from stem.
C. murale Kenya, Tanzania. Locally common.	5 Black	Base wedge-shaped. 5—15 coarse, irregular teeth directed forwards.	Leafy almost to top. Branches short, crowded with dense flower clusters.	Seeds sharply keeled.
C. fasciculosum Kenya, northern Tanzania. Locally common.	5 Black	Base rounded. Numerous, irregular, coarse, sharply pointed teeth directed outwards.	As *C. murale*	Seeds without distinct keel.

B. Plants not mealy, but with yellow glands, leaves aromatic.

Species	No. of stamens and seed colour	Leaf shape and margin	Inflorescence	Other characters
C. ambrosioides Kenya, Uganda, Tanzania. Locally common.	4—5 Deep red-brown or blackish	Long and narrow. Entire or with regular, shallow teeth or lobes.	Leafy, much branched. Small, dense flower clusters sessile in leaf or bract axils, not crowded.	3—4 stigmas (2 in other species)
C. procerum Kenya, Uganda, Tanzania. Locally common.	1—2 Black	Apex acute. 3—5 sharp toothed lobes, lower almost reaching midrib, upper shallow.	Flower clusters branched regularly, arranged in continuous cylindrical head.	Often red tinged. Generally larger and more branches than *C. schraderianum*
C. schraderanum Kenya, Uganda, Tanzania. One of commonest.	1—2 Black	Apex obtuse. 3—5 blunt lobes, all nearly reaching midrib.	As *C. procerum*	
C. carinatum Kenya. Not common	1 Deep red-brown	Small. 2—4 coarse teeth or lobes.	Small, rounded flower cluster, sessile in axils of most leaves.	Branched at base. Sepals broadly keeled.
C. pumilio Kenya. Locally common.	1 Deep red-brown	As *C. carinatum*	As *C. carinatum*	Slender branches from base. Sepals not keeled.

COMMELINACEAE

Commelina benghalensis L.   Wandering Jew

C. africana L.

DESCRIPTION

Somewhat fleshy, branched, sprawling, perennial herbs, rooting at the nodes. Leaves alternate, simple and parallel-veined, the blade ovate, with an entire margin, contracted at the base into a narrow, stalk-like portion and a sheath enclosing the stem. There are 1—3 showy flowers at the ends of the branches and the flower stalks are wholly or partially concealed within a broad, flattened spathe. The flowers consist of 3 small sepals, 3 petals, 2 of which are large and of delicate texture while the third is much smaller, and 6 stamens, 4 with yellow and 2 with blue anthers. The ovary develops into a capsule containing a small number of seeds.

Of the blue-flowered species, *C. benghalensis,* the commonest, is distinguished by the flower stalk, which does not extend beyond the spathe, by the spathe margins being joined for part of their length and by the rusty hairs on the leaf sheath. *C. latifolia* A. Rich. and *C. diffusa* Burm.f. also occur as weeds, the former in most parts of East Africa, the latter being especially important in Uganda. *C. africana* is the only common yellow-flowered species.

Distribution and importance

C. benghalensis occurs widely in the tropics, *C. africana* in many African countries, and both are common throughout East Africa from sea level to 2000 —2500 m. They grow in a wide range of situations in grassland or arable crops, especially in moister areas, and are often common in coffee. Their scrambling habit leads to the smothering of such low growing crops as groundnuts. *C. benghalensis* is particularly difficult to control by cultivation, partly because broken pieces of stem readily take root again and partly because underground stems with pale reduced leaves and flowers are often produced.

Methods of control

Commelina species are persistent weeds and not easy to control. Hormones such as 2,4-D can be effective, but only on small seedlings. Paraquat also is only effective in the early stages but glyphosate kills older plants and has given good results prior to sowing wheat in Tanzania. Bentazon is one of the more active contact herbicides on *Commelina* and can be used in cereal crops and soya bean. Metribuzin has given good results in a number of crops, but most residual chemicals are unreliable.

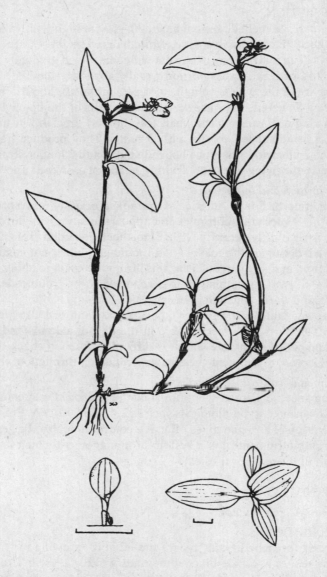

Figure 58. Commelina benghalensis

COMPOSITAE
Acanthospermum hispidum DC.   Starbur
DESCRIPTION
A regularly branched, annual herb, 30 —60 cm high with yellow flowers
and the stem covered in coarse, white hairs. The leaves are simple and
opposite, up to 8 cm long × 2.5 cm wide, narrowed gradually to a sessile
base and with the upper portion usually shallowly toothed. The flower
heads are of the typical composite type, but small, about 5 mm across,
and few-flowered. They are sessile in the axils of the upper leaves. The
outer ring of flowers in the head (ray florets) are 7 —8 in number and
have a strap-shaped corolla with 3 teeth. They produce hard, single-
seeded fruits with numerous hooked spines and 2 longer straight spines
at the apex, giving the whole fruiting head a star-shaped appearance.

Distribution and importance
Originating in South America *Acanthospermum* is now spread widely
over the tropics and subtropics and appears to have been introduced into
East Africa fairly recently. It has not been recorded from Uganda; in
Kenya it occurs on the coast and in Ukambani, but is not common, while
it is found in many parts of Tanzania, from the coast to about 1200 m. In
the Morogoro-Kilosa district it is reported to be troublesome in cultivation and waste land and to be increasing rapidly.

The weed infests the poorer types of grassland and all types of arable
crop. The burrs are objectionable in the hair of animals and also assist
seed dispersal. In arable land it is difficult to keep in check through the
life of a crop, as the seed germinates irregularly throughout the season.

Methods of control
Young seedlings are susceptible to contact herbicides such as bentazon
and to hormone-type chemicals such as 2,4-D and MCPA. Pre-emergence
application of atrazine gives effective control in maize. In grassland the
best long-term control is a system of management which increases the
vigour of the pasture species.

COMPOSITAE
Ageratum conyzoides L.   Goat Weed
DESCRIPTION
An erect, branched, softly hairy, annual herb, growing to 1 m, with opposite leaves and pale blue, sometimes white, flowers. The leaves are
stalked and ovate, up to 8 cm long × 5 cm wide, with a more or less
pointed tip and the margin regularly serrated with blunt teeth. The

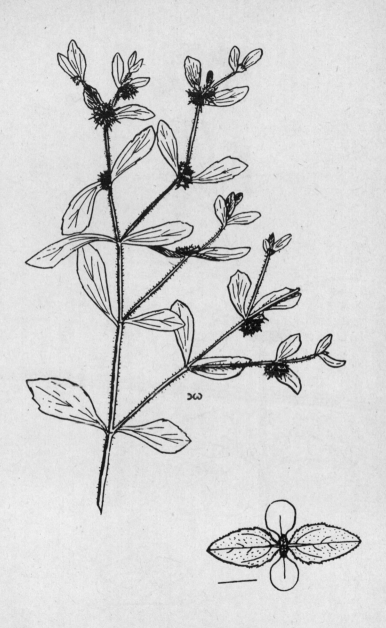

Figure 59. Acanthospermum hispidum

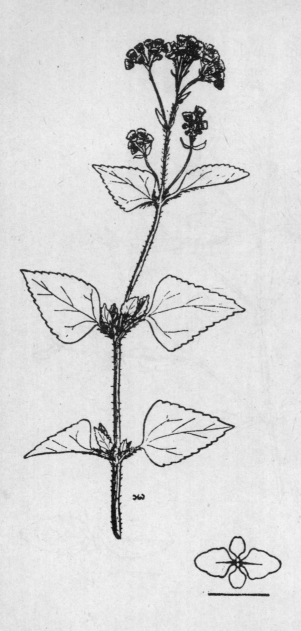

Figure 60. Ageratum conyzoides

inflorescence is terminal and made up of several branches, each bearing a number of flower heads, which are arranged in a showy, more or less flat-topped cluster. Individual flower heads are 5 mm across and made up of a large number of tubular flowers surrounded by 2 or 3 rows of narrow, pointed bracts with membranous margins. The one-seeded fruits are black, ribbed and crowned with 5 white bristles.

Distribution and importance
A common arable weed in many tropical and sub-tropical countries, in some areas occurring predominantly in the white-flowered form. It is widespread in East Africa from the coast to about 3000 m, especially in the higher rainfall districts. It is recorded as a particular problem in the Nyanza district of Kenya.

Ageratum occurs in a wide range of arable crops, in waste land and to some extent in grassland. It produces seed in large quantities and dense stands of seedlings often develop. It has been suspected of causing poisoning in cattle, but feeding tests in Australia have not confirmed this.

Methods of control
Seedlings and young plants are readily controlled by 2,4-D, MCPA and other growth-regulating type herbicides at the normal doses employed in cereal crops. Various residual chemicals are also reported to give control including such materials as atrazine, prometryne, metribuzin, diuron, linuron, chloroxuron and alachlor.

COMPOSITAE

Bidens pilosa L.	Blackjack
B. biternata l. (Lour.) Merr. & Sherff	Yellow-flowered Blackjack
B. schimperi Walp.	Yellow-flowered Blackjack
B. steppia (Steetz) Sherff	

DESCRIPTION
Bidens pilosa is an erect, branched, annual herb growing to about 60 cm high with opposite leaves, white flowers, and a 4-angled stem. The leaves are made up of 3 (sometimes 5) stalked, ovate leaflets with acute tips and sharply serrated margins, the terminal leaflet being larger than the lateral ones. Flower heads are about 1 cm across, borne on long stalks and arranged in a branched, loose, terminal inflorescence. In fruit the heads become elongated and widened towards the top. There are about 5 broad, white ray florets (these are sometimes absent) 5 mm long and 3-lobed at the tip, and numerous yellow, tubular disc florets. The bracts surrounding the flower heads are in two rows and the outer ones

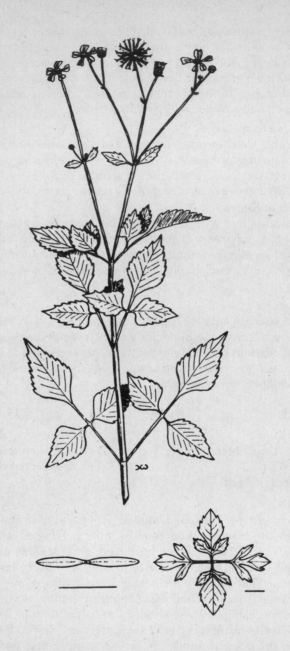

Figure 61. Bidens pilosa

are broadened towards the tip. Fruits are black, narrow, 5—10 mm long and bear 2—3 stout, hooked bristles.

B. biternata is very similar to the above with flowers the same size, but the ray florets are yellow and the leaves usually have 5—9 leaflets arranged pinnately, the lower ones often themselves divided into 3. The outer bracts round the flower heads are not broadened at the tip. The fruits are longer, up to 20 mm, and bear 3—5 hooked bristles.

B. schimperi is another yellow-flowered species with larger flowers. The ray florets are more numerous and 10 mm or more long. It has roughly hairy leaves consisting of 3 leaflets, usually themselves deeply divided into 3, and the margins are marked with a line of black dots. Fruits are 5—10 mm long with 2 hooked bristles.

B. steppia has flowers like *B. schimperi,* and is distinguished by having the two bristles on the fruits smooth or with merely a few forward-pointing hairs, not hooked as in *B. schimperi.*

Distribution and importance

B. pilosa is a cosmopolitan weed of warmer areas, the other species being less widely distributed. *B. pilosa* is also the commonest species in East Africa and occurs in all areas, often as one of the most important annual weeds. *B. biternata* and *B. schimperi* are also widespread in East Africa, but less frequently recorded and not generally present in such large numbers. *B. steppia* sometimes occurs as a weed in Tanzania and Uganda.

All the species produce large quantities of seed, which can germinate immediately after being shed. With *B. pilosa* it has been shown that, if conditions are not suitable for germination for about 8 weeks, the seed becomes dormant and can remain viable for a considerable period. Besides the competitive effects of the weeds on crops, their barbed seeds can be troublesome in their ability to adhere to and penetrate the hair of animals, clothing, skin, etc. In Zimbabwe *Bidens* species have been shown to act as alternate hosts of various eelworm pests of crops.

Methods of control

Information on susceptibility to herbicides is largely confined to *B. pilosa,* but there is no evidence that the other species react very differently. Good control can be obtained in cereal crops with 2,4-D or MCPA while the contact herbicides, including bromoxynil, ioxynil and bentazon are also effective on this species, effective residual herbicides include atrazine, metribuzin, linuron etc. *B. pilosa* has often been observed to appear in large numbers in coffee when infestations of perennial grass weeds have been eradicated by treatment with dalapon.

COMPOSITAE

Carduus kikuyorum R. E. Fries   Kikuyu Thistle

C. chamaecephalus (Vatke) Oliv. & Hiern   Stemless Thistle
  (= *C. theodori* R. E. Fries)

DESCRIPTION

C. kikuyorum is an erect, prickly, little branched perennial, growing to a

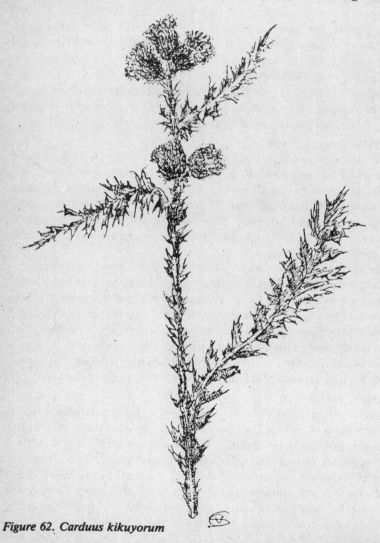

Figure 62. Carduus kikuyorum

height of up to 2 m, with pale purple or whitish flowers and a deeply grooved stem. The alternately arranged, roughly hairy leaves are up to 40 cm long × 10 cm wide and deeply divided into pinnate lobes, the lobes themselves deeply cut into spine-tipped teeth and with a spiny margin. The basal leaves are stalked, the upper leaves narrowed gradually to the base with spiny wings continuing down the stem. The sessile flower heads are 2 —2.5 cm across, and borne in clusters, 2 —5 together at the ends of the main stem and branches. They consist entirely of tubular florets which are surrounded by several rows of narrow, spine-tipped bracts. The wind-distributed fruits are crowned with a ring of numerous, stiff, whitish hairs about 2 cm long.

   C. chamaecephalus is similar in its flowers and prickly leaves, but has no stem, so that the flowers are closely appressed to the centre of a dense rosette of leaves at ground level. Flower heads are 4 —5 cm across and the leaves differ from those of *C. kikuyorum* in being without hairs.

Distribution and importance
These two indigenous thistles are found mostly in grassland at altitudes above about 2000 m. They occur in parts of Uganda and Tanzania, but are only of significance as weeds in the highlands of Kenya, being recorded as of local importance in the Laikipia, Uasin Gishu and Elgeyo-Marakwet districts. They become most numerous in low-lying, seasonally waterlogged areas and dense infestations can greatly reduce the grazing potential over limited areas.

Methods of control
There is little information on the reaction of these weeds to herbicides, but temperate *Carduus* species are relatively susceptible to the action of MCPA and 2,4-D and it would be expected that the above-ground parts of the East African species would also be readily killed. To what extent the underground parts would be affected is not known, but resprouting might well be delayed or prevented by using a formulation containing dicamba. Application of glyphosate by wiper bar is also a promising technique for controlling tall-growing thistles.

COMPOSITAE

Centaurea melitensis L.   Maltese Thistle
(*C. muricata* L.) see *Volutaria muricata*
(*C. lippii* L.) see *Volutaria lippii*

DESCRIPTION
C. melitensis is an erect, much branched, annual herb reaching a height of up to 1 m, with alternate leaves and yellow flower heads surrounded

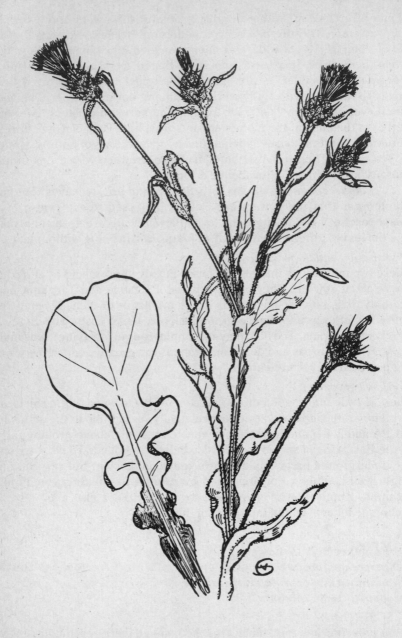

Figure 63. Centaurea melitensis

by spiny bracts. The stems are clothed with cottony hairs and are broadly winged. The leaves are also cottony on both sides, the lower ones deeply and pinnately cut into narrow lobes, the upper lanceolate and more or less entire. The leaves are sessile and the leaf blade is connected with the wings running down the stem. Flower heads are numerous, 1 cm across and borne singly on very short stalks. Florets are all tubular and similar in size. They are surrounded by several rows of overlapping, cottony bracts, each of which terminates in a stiff brownish spine about 5 mm long and has shorter lateral spines in its lower half. The seeds are pale grey, striped with white and crowned with short, whitish hairs.

Distribution and importance

C. melitensis is native to the Mediterranean region, from where it has been introduced into various other parts of the world, including the Kenya highlands. It is recorded as a locally serious problem in small-grain crops, maize and grass leys in the Uasin Gishu, Laikipia and Molo districts. Several species of *Centaurea* are known to contain poisonous alkaloids and *C. melitensis* has been suspected of poisoning livestock in Australia.

Methods of control

In Australia 2,4-D is recommended for controlling *C. melitensis,* applied at a rate of 1 kg per ha to seedlings or small rosettes. The plant becomes more difficult to kill as it becomes older and, once the flower stem has started to elongate, higher rates or a stronger chemical, such as a dicamba formulation, are needed.

COMPOSITAE

Conyza bonariensis (L.) Cronq.   Fleabane
   (= *Erigeron bonariensis* L. = *E. linifolius* Willd.)
C. floribunda H.B.K.
   (= *Erigeron floribundus* (H.B.K.) Sch. Bip.)

DESCRIPTION

These two species are separated from those in the next section as they are distinctive, both in appearance and in their status as weeds, and were until recently in the genus *Erigeron.*

The species are very similar in general appearance, being annual herbs with stiffly erect, leafy stems, which arise from a basal rosette of leaves to a height of up to 1.2 m and with long, narrow, alternate leaves and numerous small, yellowish flower heads. The stem is ribbed, hairy, tough and woody at the base. The leaves are unstalked, up to 8 cm long × 1 cm wide and have a smooth margin. The flower heads are borne on

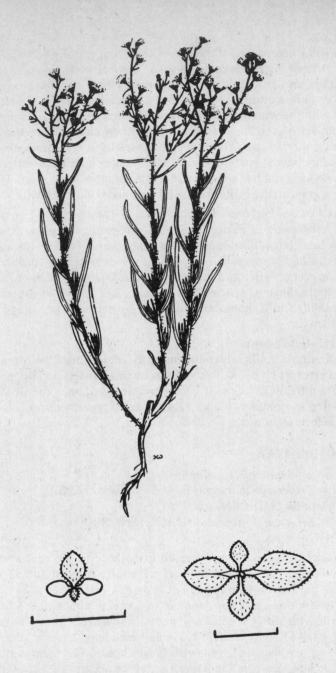

Figure 64. Conyza bonariensis

stalks about 1 cm long, in an elongated, terminal inflorescence. Individual heads are 5 —8 mm across and made up of narrow, tubular florets, with numerous long hairs arising from the ovary. They are surrounded by several rows of narrow, pointed bracts which turn downwards after the fruits have been shed.

C. bonariensis has long inflorescence branches, equalling or exceeding the main stem, the hairs attached to the fruit are white or pinkish and the inner surface of the bracts surrounding the flower heads is white when the bracts are old and turned downwards.

In *C. floribunda* the inflorescence branches are shorter and far from equalling the height of the main stem. The hairs on the fruit are straw-coloured and the inner surface of the old bracts is brown.

Distribution and importance
Both species appear to have originated in South America, but have now been distributed throughout the tropics and occur widely in all the East African countries, from sea-level to at least 2500 m. Both are very common and they are found in all types of crops, in waste land and in gardens. They are often noted as weeds of coffee and commonly occur in large numbers in fallow land. Seed production is copious and the seeds can germinate immediately after being shed, so that dense infestations can develop in a relatively short time.

Methods of control
In the early stages of growth the plants can be controlled readily by cultivation, but later, when the base of the stem becomes tough, they are much more difficult to deal with.

Similarly, with herbicides of the 2,4-D and MCPA type they can be readily killed as young seedlings, but soon develop resistance. Formulations containing dicamba, picloram or clopyralid are more effective on older plants and amitrole or glyphosate will also give control.

Among the residual herbicides substituted ureas (diuron, linuron, metoxuron etc.) and triazines (atrazine, prometryne, metribuzin etc.) are effective as pre- or early post-emergence treatments.

COMPOSITAE

Conyza stricta Willd.
C. aegyptiaca (L.) Ait.
C. schimperi A. Rich.
C. steudelii A. Rich. (= *C. volkensii* O. Hoffm.)
DESCRIPTION
These four *Conyza* species are spreading, much branched annual or

short-lived perennial herbs, with alternate leaves and cream or pale yellow flower heads, densely clustered at the ends of the branches to form more or less flat inflorescences. The florets making up the flower heads are all tubular and are only slightly longer than the surrounding bracts. The seeds are crowned with a pappus of long brownish or whitish hairs.

C. stricta grows to a height of about 60 cm. The coarsely hairy leaves are sessile and up to 2.5 cm long, narrowly spoon-shaped, undivided or cut into irregular, pinnate lobes, and have a base which clasps the stem. The flower heads are about 5 mm diameter and the length of the longer surrounding bracts averages 2.5 mm.

C. aegyptiaca is usually a larger plant reaching 1 m high. It also has sessile leaves, which are up to 1 cm long × 2.5 cm wide and very variable in shape, entire or more often pinnately lobed, coarsely toothed and with the basal lobes either clasping the stem or projecting on either side. The flower heads are 8 —12 mm across and the longer surrounding bracts 7 mm.

C. schimperi is generally similar to *C. aegyptiaca,* but the leaves are distinctly thicker, pinnately and regularly divided into narrow lobes, rounded at the tip, and the flower heads are smaller. The longer bracts average 5 mm.

C. steudelii has flower heads and bracts the same size as *C. schimperi.* Unlike the other species, however, the leaves are sharply toothed, pointed at the apex and narrowed into a winged stalk, which is expanded at the base into two lobes which more or less clasp the stem.

Distribution and importance

All four species are indigenous plants and *C. stricta* is the most widely distributed and the commonest in the three East African countries. *C. schimperi* and *C. steudelii* are frequently found in the Kenya highlands and occur less often in Uganda and Tanzania, the former being the more frequent, while *C. aegyptiaca* is more common in Tanzania and Kenya. *Conyza* species are reported as particularly common weeds in the Machakos district of Kenya. They are principally weeds of waste ground, but also occur quite frequently in arable land.

Methods of control

There is little information on the susceptibility of these *Conyza* species to herbicides but *C. stricta* appears to show some susceptibility to growth-regulator type herbicides. It is probable that young seedlings would be adequately controlled by MCPA or 2,4-D but well established plants are likely to require more than one treatment or the use of a stronger chemical, such as a formulation containing dicamba.

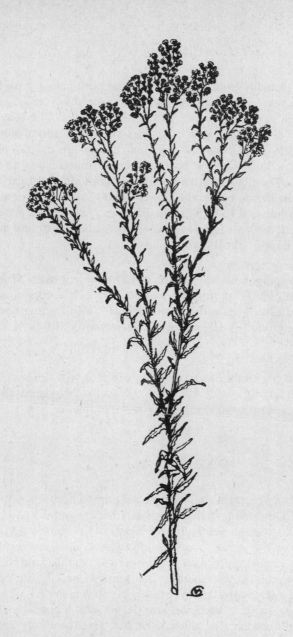

Figure 65. Conyza stricta

COMPOSITAE

Crassocephalum rubens (Jacq.) S. Moore

DESCRIPTION

This is a soft-stemmed, erect, branched annual herb growing up to 1 m tall with purple, somewhat nodding, solitary flower heads on long stalks. The stems are ridged and covered in fine, short hairs. The leaves are alternate, those at the base with a distinct stalk, the upper ones with basal lobes which clasp the stem. The leaf blades are cut deeply into several pairs of pointed lobes below the larger terminal lobe and have a toothed margin. The flower stalks are 10—20 cm long and bear a few, scattered, narrow bracts below the flower head. The heads themselves are oblong in outline and 10—15 mm in diameter. The tubular florets are purple, surrounded by a single row of narrow, pointed bracts and develop a copious pappus of silky white hairs.

Distribution and importance

C. rubens is an indigenous species distributed widely over Africa and common at a wide range of altitudes in East Africa. It occurs most commonly as a weed associated with cultivation but is also frequent on roadsides and waste ground. There are numerous reports of it as a weed in maize crops.

Methods of control

In common with most other annual composite weeds *C. rubens* can be readily controlled with a wide range of contact, translocated and residual herbicides and presents no special control problems.

COMPOSITAE

Dichrocephala integrifolia O.Kuntze

DESCRIPTION

D. integrifolia is a vigorous, much branched annual herb, growing 60—100 cm tall, with alternate, dull green, hairy leaves and a leafy inflorescence of small, greenish-yellow flower heads. The stem is ridged and hairy, the hairs on the young shoots forming a grey, woolly covering. The leaves are up to 15 cm long × 6 cm wide with a long stalk, and the blade is divided to the midrib into usually two pairs of small, sub-opposite basal lobes and a large, triangular, bluntly pointed terminal lobe which makes up half or more of the total length, all with a coarsely toothed margin. The inflorescence is much branched and the many spherical flower heads are 3—4 mm in diameter and borne on stalks about 1 cm long. Each head is made up of numerous small, tubular florets, sur-

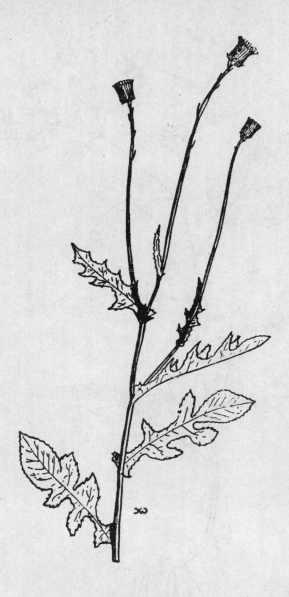

Figure 66. Crassocephalum rubens

Figure 67. Dichrocephala integrifolia

rounded by inconspicuous ray florets and there are no hairs or scales on the flattened, single-seeded fruits.

Distribution and importance

This is a native species common in East Africa at medium altitudes. It is noted particularly as a weed in Western Kenya occurring in coffee and various other crops. In the Limuru district it is recorded as a weed of potatoes.

Methods of control

No information is available specifically on *D. integrifolia* but, in common with other annual weeds belonging to this family, it would be expected to be susceptible to a wide range of foliar and residual herbicides.

COMPOSITAE

Galinsoga parviflora Cav.   Macdonaldi, Gallant Soldier

DESCRIPTION

A much branched, annual herb of soft growth, reaching a height of about 60 cm with opposite leaves and small white flowers. The leaves are slightly hairy and simple, with a short but distinct stalk and an ovate blade up to 5 cm long × 4 cm wide, having three conspicuous veins converging at the base. The leaf margin is shallowly indented with small teeth. The flower heads are 5 mm across on stalks 1—2 cm long, and are arranged to form a regularly branched, leafy and rather loose inflorescence at the apex and in the axils of the upper leaves. Each head is made up of about 5 three-lobed ray florets, only slightly longer than the surrounding bracts, and numerous, tubular, yellow disc florets. The latter bear a ring of flat, silvery, fringed scales about the same length as the corolla, and these persist as a crown at the apex of the black seeds.

Distribution and importance

Originally a native of South America, *Galinsoga* is now widespread in tropical, sub-tropical and temperate regions, being one of the commonest arable weeds in many areas. It occurs on agricultural land throughout East Africa in all types of crops, on waste land and in gardens. It produces particularly large quantities of seed which germinate very readily. Viable seed can be produced by plants only a few inches high and stands of the weed are often dense enough to exclude all other species.

Methods of control

Young plants are susceptible to MCPA, 2,4-D and other growth-regulator type herbicides as well as to most contact chemicals. Among the

138

residual herbicides such chemicals as atrazine, metribuzin, diuron, linuron, metoxuron and alachlor give effective control but *Galinsoga* is relatively resistant to chemicals of the carbamate and dinitroaniline groups.

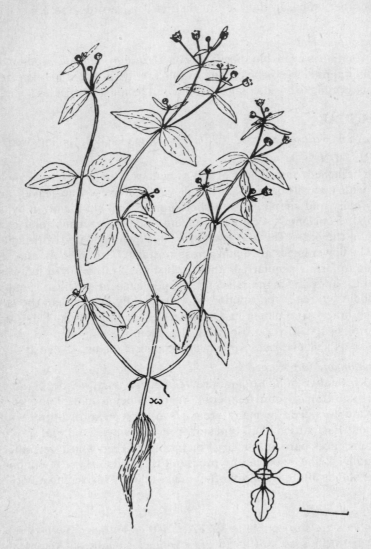

Figure 68. Galinsoga parviflora

COMPOSITAE

Gnaphalium luteo-album L.   Cudweed

DESCRIPTION

An annual herb growing to about 50 cm with spirally arranged leaves and yellowish flower heads, the stem leaves having a dense, woolly covering of white hairs. The plant branches near the base, the branches at first radiating out horizontally and then becoming erect. There is little further branching below the apical inflorescence. The basal leaves are oblanceolate and blunt; stem leaves are narrower and more pointed, up to 8 cm long, with a more or less wavy margin, and narrowed gradually to the stem-clasping base. Individual flower heads are 3 mm across but are crowded densely, 5 —10 together, in leafless, globular clusters at the ends of the inflorescence branches. The florets comprising the heads are all tubular and the same length as the shining, straw-coloured bracts which surround them in several overlapping rows.

Distribution and importance

A weed of wide distribution in Africa, other tropical regions and the warmer parts of Europe. It is common throughout East Africa in arable land, gardens, poor pastures and waste ground, but, although generally distributed, does not appear to be of any outstanding importance.

Methods of control

Variously reported as being moderately susceptible and only slightly susceptible to 2,4-D and MCPA, it is probable that complete control can only be obtained by treatment in the young seedling stage. Formulations containing dicamba are more effective and amitrole or glyphosate can be used as non-selective treatments to kill older plants. *G. luteo-album* is susceptible to pre-emergence treatment with linuron and monolinuron and related species are readily controlled with atrazine, diuron and several other soil-applied herbicides.

COMPOSITAE

Gutenbergia cordifolia Oliv.
(= *Erlangea cordifolia* (Oliv.) S. Moore)
G. marginata Oliv. & Hiern
(= *E. marginata* (Oliv. & Hiern) S. Moore)

DESCRIPTION

G. cordifolia is an erect, branched annual herb, growing to 0.6 —1 m high, with simple, alternate leaves and numerous heads of bright purple flowers. The stems and undersides of the leaves are covered with silvery

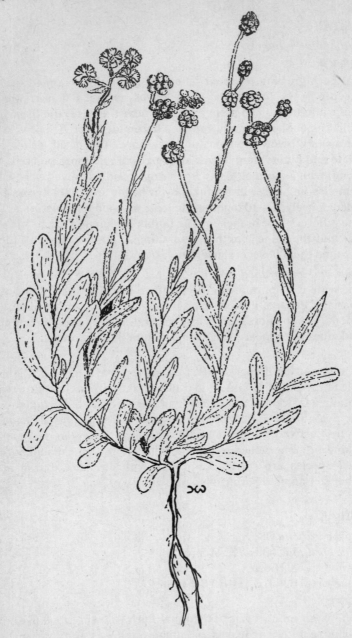

Figure 69. Gnaphalium luteo-album

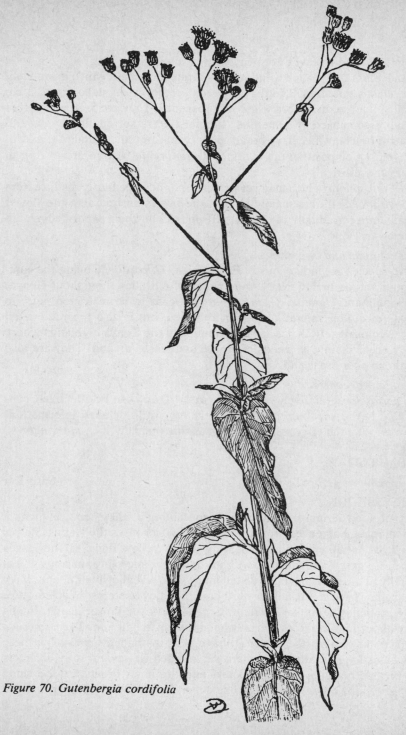

Figure 70. Gutenbergia cordifolia

hairs. The stem leaves are ovate, up to about 8 cm long × 4 cm wide with a blunt tip, shallowly indented margin and a base which is stalkless and clasps the stem. The flower heads are borne on stalks up to 2 cm long in a loose, much branched inflorescence. They are 5 —8 mm across and are composed of numerous, tubular florets surrounded by several rows of bracts, which have broad, papery margins and rounded tips. Unlike most Compositae, the fruits of *Gutenbergia* lack an apical ring of hairs or scales.

    G. marginata (which may yet prove not to be specifically distinct from *G. cordifolia*) differs mainly in that the bracts surrounding the flower heads taper gradually to sharp, stiff points and their papery edges are narrower.

Distribution and importance
Both species are indigenous in East Africa, *G. cordifolia* being the commoner and occurring over a wide range of altitudes throughout Kenya, Tanzania and Uganda. *G. marginata* appears to be restricted more to areas of higher rainfall at higher altitudes, but is also recorded in all three countries. Both species are common in the Kenya highlands where they occur mainly as weeds of arable and waste ground. They are also found in poor grassland.

Methods of control
G. cordifolia is relatively susceptible to 2,4-D and can be effectively controlled in the seedling stage. There is very little detailed information, however, on the effects of herbicides on this weed or on *G. marginata*.

COMPOSITAE

Hypochoeris glabra L.   Cat's Ear

DESCRIPTION
A taprooted annual or biennial herb containing a milky juice, and with a basal rosette of leaves, from the centre of which rises the leafless flower stalk, about 30 cm high and bearing bright yellow flowers. The leaves are pale green and smooth, without hairs, oblanceolate, narrowed gradually to the base and up to 25 cm long, with a shallowly or deeply wavy margin. There are usually several branched flower stems to each rosette, the branches being somewhat enlarged below the flower heads. Heads are about 15 mm across and only open fully in bright sun. The numerous florets making up the heads all have a strap-shaped terminal portion, twice as long as broad. They are surrounded by several rows of bracts with pale margins and dark points and the inner bracts are the same length as the outer florets. The one-seeded fruits are reddish brown, the

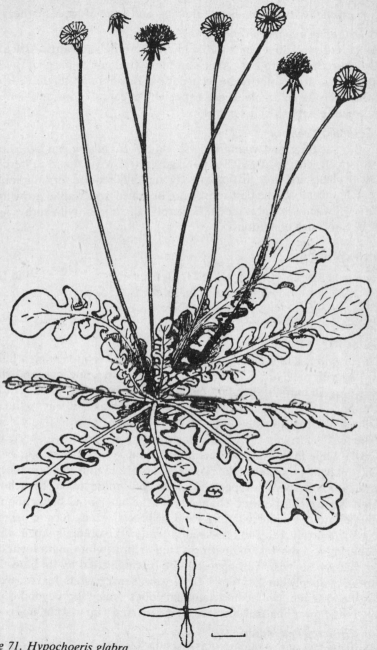

Figure 71. Hypochoeris glabra

central ones with a long terminal point, and bear a ring of feathery hairs.

Distribution and importance

A plant of sand dunes and arable land in Europe and North Africa, probably introduced into East Africa and now common in parts of the Kenya highlands. It has not been recorded in Tanzania or Uganda. It occurs principally as an arable weed, especially in cereal crops and sometimes becomes established in pastures.

Methods of control

In the seedling and young rosette stages *H. glabra* can be controlled effectively with 2,4-D, MCPA or other growth-regulator type herbicides. Older plants are more difficult to kill and may require formulations containing dicamba. The best results are obtained under good growing conditions, when the soil is moist. Control is often poor with sprays applied under hot, dry conditions.

COMPOSITAE

Launaea cornuta (Oliv. & Hiern) C. Jeffrey   Wild Lettuce
(= *Lactuca taraxacifolia* sensu Ivens ed. 1, = *Sonchus exauriculatus* (Oliv. & Hiern) O. Hoffm.)
Lactuca capensis Thunb.

DESCRIPTION

L. cornuta is a tall, erect, glabrous, perennial herb reaching a height of 1.2 m with pale yellow flowers and an extensive underground rhizome system. All parts of the plant contain a copious, milky latex. The leaves are up to 15 cm long and pinnately cut as much as half way to the midrib into sharp, backwardly-directed teeth. The margin is finely serrated. The leaves forming the basal rosette are narrowed into a short, winged stalk, while the alternately arranged stem leaves are sessile, with the base clasping the stem. The flower heads are 15 mm across on very short stalks, rather widely separated along the inflorescence branches. The florets are all of the strap-shaped type, the outer rows being about twice the length of the surrounding rows of bracts, which form a narrow, cylindrical involucre. The one-seeded fruits are spindle-shaped, the apex tapering to a short point bearing a ring of long, white, unbranched hairs.

Lactuca capensis can generally be distinguished by its blue flowers, the less deeply cut leaf teeth, the entire, sessile, stem leaves, extended at the base into two acute triangular lobes projecting beyond the stem, and the flower heads, which are borne on stalks up to 4 cm long.

Distribution and importance

Both species occur widely in Africa and are distributed in many parts of

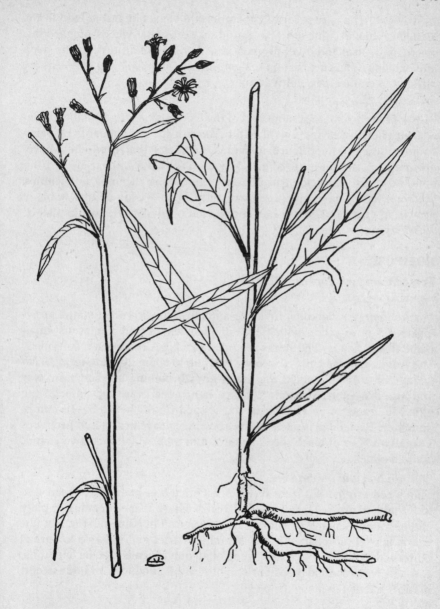

Figure 72. Launaea cornuta

East Africa. They are recorded as important weeds in arable land in the Kenya highlands, but are also found as occasional weeds of gardens, waste land, sisal and other crops at a wide range of altitudes. The deep and extensive rhizome system makes *Lactuca* and *Launaea* species very difficult to eradicate by cultivation.

Methods of control
Most of the rhizomatous members of the Compositae are difficult to control with herbicides and *Lactuca* and *Launaea* species appear to be particularly difficult. Experience in Kenya suggests that even the above-ground parts are but little affected by 2,4-D or MCPA. In trials in Ethiopia, however, herbicide formulations containing dicamba or picloram are reported to have given good results. With weeds of this habit of growth in grassland, wiper-bar application of glyphosate would also be worth trying.

COMPOSITAE

Parthenium hysterophorus L.

DESCRIPTION

Parthenium is a vigorous, invasive annual herb with erect stems growing to 1.5 m or more, with finely divided leaves and numerous small white flower heads. The stems are ridged and covered in fine, soft hairs. The leaves are also hairy, the lower ones up to about 20 cm long × 10 cm wide, with a distinct stalk, the upper ones becoming smaller, narrower and less dissected. The inflorescence is a much branched panicle, the ultimate branches of which bear flat topped flower heads 5 —10 mm in diameter. Each head is enclosed by about 5 concave bracts and produces a small number of black seeds about 2 mm long which bear two round, whitish scales.

Distribution and importance
This weed originating from tropical America has been introduced into India and, more recently, into Australia where it is spreading rapidly and becoming a major problem. It has appeared in Kenya in the last few years, is well established in the Kiambu district and is liable to spread more widely. It is a noxious weed, not only because of its vigorous, competitive growth in crops but also because it produces an intense dermatitic reaction in many people.

Methods of control
Australian work suggests that *Parthenium* is not difficult to control with growth-regulator type herbicides. 2,4-D is effective on seedlings and dicamba formulations on older plants. This weed has shown itself cap-

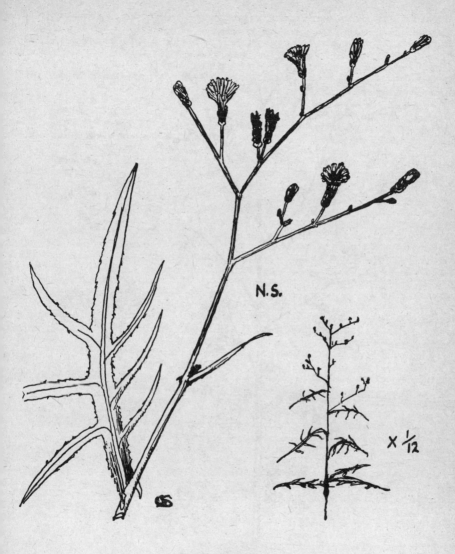

N.S.

X $\frac{1}{12}$

Figure 61. Lactuca capensis

Figure 73. Lactuca capensis

Figure 74. Parthenium hysterophorus

able of producing very large numbers of seeds in a short time. The light seed is readily distributed by wind, by passing vehicles along roadways etc., and once an infestation has become established, some residual herbicide treatment is generally needed to deal with successive flushes of germinating seedlings. Atrazine is effective as a residual treatment and hexazinone also gives good control. It is desirable that all possible measures should be taken to prevent any further spread of this weed in East Africa.

COMPOSITAE

Schkuhria pinnata (Lam.) Thell.   Dwarf Marigold
(= *S. isopappa* Benth.)

DESCRIPTION

A much branched annual herb up to 50 cm high with numerous, small, yellow flowers and finely divided, alternate leaves. The stem and branches are slender and ribbed. The leaves are dotted with glands and up to 8 or 10 cm long, pinnately divided into very narrow segments, usually about 3 pairs, with the lower segments often themselves pinnately divided. The flower heads are 5 mm long and rather less than 5 mm across, borne on long, slender stalks and arranged in a loose, branched, leafy, terminal inflorescence. The flower heads consist of 3 —9 central, tubular florets and an outer ring of a small number of strap-shaped florets. These outer florets are short and inconspicuous and only slightly exceed the surrounding row of bracts, about 6 in number. The single-seeded fruits are crowned with a ring of 8 brownish scales.

Distribution and importance

A weed introduced from South America into various parts of Africa. It has found its way into the Kenya highlands and now occurs in several places from Nairobi to Kisumu. In Tanzania it has been recorded from the Moshi district, but is not common, and does not appear to have entered Uganda. It is principally a weed of arable crops and fallow land. Seeds can remain dormant for considerable periods.

Methods of control

Young plants are sensitive to 2,4-D and MCPA, while pre-emergence treatment controls germinating seedlings. The effectiveness of most other chemicals of the growth-regulator type is similar to that of 2,4-D and MCPA, but MCPB appears to be slightly less effective. Little information on residual herbicides is available but *Schkuhria* would be expected to be susceptible to chemicals of the triazine and substituted urea groups.

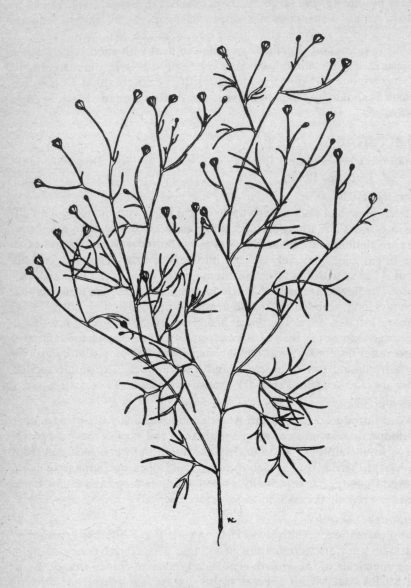

Figure 75. Schkuhria pinnata

COMPOSITAE

Senecio discifolius Oliv. Ragwort
S. abyssinicus A.Rich.
S. vulgaris L. Grounds

DESCRIPTION

These three annual *Senecio* species are erect, rather weak-stemmed herbs, growing up to 50 cm high with alternate, simple leaves and yellow flowers. *S. discifolius* has leaves with a more or less circular blade up to 5 cm across and abruptly narrowed into a winged stalk. The margin is shallowly toothed and the 3 terminal teeth are larger than the rest. The flower heads are about 2.5 cm across and borne singly or up to 3 together on stalks 7 —17 cm long, at the ends of the branches. They consist of numerous tubular florets and an outer ring of much longer, spreading, strap-shaped ray florets. The surrounding bracts form a single row. The fruits are cylindrical and crowned with numerous, long, white hairs.

  S. abyssinicus is generally similar, but can be distinguished by its very slender, cylindrical flower heads, without spreading ray florets. The diameter of the ring of bracts surrounding the flower heads does not exceed 3 mm, by comparison with *S. discifolius* where it is generally 6 mm or more.

  S. vulgaris also differs most obviously in the flower heads, which lack ray florets and are thus smaller and less showy than in *S. discifolius*. The flower heads have very short stalks and are borne in dense, more or less flat-topped clusters. The leaves are deeply and pinnately divided into broad, blunt-ended, irregularly toothed lobes. The lower leaves are shortly stalked, the upper ones sessile and partly clasping the stem.

Distribution and importance

S. discifolius and *S. abyssinicus* are indigenous species. *S. discifolius* is common and widely distributed through East Africa. It is especially common in the Machakos district of Kenya. *S. abyssinicus* appears to be much commoner in Tanzania than in Kenya or Uganda. *S. vulgaris* has been introduced from Europe and is common in parts of the Kenya highlands. It has not so far been recorded in Uganda or Tanzania. All three species are typical arable weeds and *S. discifolius* is also common in waste land. *S. vulgaris* occurs principally in cereal crops, grows very rapidly and seeds very profusely.

Methods of control

The effects of herbicides on *S. vulgaris* are well documented from European experience. In the young seedling stage reasonable control can be obtained with 2,4-D, MCPA and most other growth-regulator type

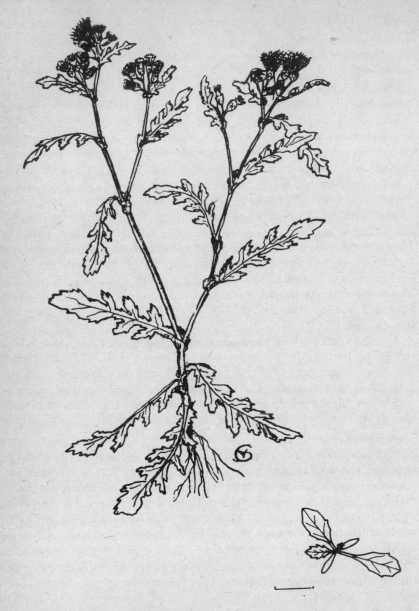

Figure 76. Senecio vulgaris

chemicals (not MCPB). There is a considerable increase in resistance with age, however, and after the 3-leaf stage, these materials have little effect. Early spraying is likely to be even more important under East African conditions. A mixture of MCPA and dicamba is more effective against this weed than other growth-regulators. Contact herbicides such as ioxynil or bentazon also give good control of young seedlings and soil-applied chemicals of the triazine and substituted urea groups are effective as pre- or early post-emergence treatments.

There is less information on *S. discifolius* and *S. abyssinicus*, but what evidence is available suggests that, in general, they behave in a similar manner to *S. vulgaris*.

COMPOSITAE

Senecio ruwenzoriensis S. Moore   Ragwort
S. moorei R.E. Fries   Mau Narok Ragwort

DESCRIPTION

These two perennial *Senecio* species present a very different weed problem from the annuals described in the previous section.

S. ruwenzoriensis is a herbaceous plant growing to a height of 50 — 60 cm from an underground system of rhizomes. The stems are erect and little branched, the leaves bluish-green, alternate and simple, up to 8 cm long × 3 cm broad, narrowed gradually to a sessile base, broadest somewhat above the middle and rounded at the tip, with a smooth margin. The yellow flower heads are borne on upright 1 —2 cm stalks to form a few-flowered, little branched, apical inflorescence. The heads are 1 —2 cm across and consist of a ring of about 8 strap-shaped florets, surrounding numerous shorter, tubular florets and surrounded by a single ring of bracts. The fruits are crowned with numerous, long, white hairs.

S. moorei, by contrast, is a tall, much branched herb with a woody base reaching 1.5 – 2 m high. The leaves are up to 10 cm long × 2 cm wide, with a sessile base or a short, winged stalk, a pointed tip and a serrated margin. When young they are covered with white, woolly hairs, but these are lost as the leaves mature. There are numerous yellow flower heads on 2 cm stalks arranged in a much branched inflorescence. The strap-shaped outer florets are about 13 in number and there are about the same number of bracts which bear black hairs and have papery margins.

Distribution and importance

The two species are indigenous and of restricted distribution, *S. ruwenzoriensis* being found at higher altitudes in Kenya, Tanzania and

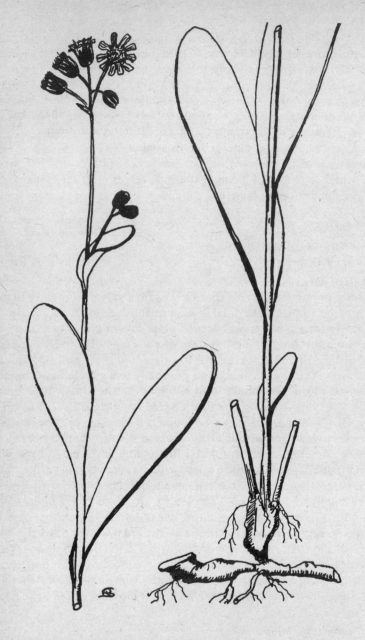

Figure 77. *Senecio ruwenzoriensis*

Uganda, *S. moorei* only in Kenya above 2000 m particularly in the Mau Narok district. Both plants are of special importance as weeds in the Kenya highlands, mainly in pastures. *S. ruwenzoriensis* is known to be poisonous to stock and *S. moorei* is suspected of being poisonous. The latter species also forms dense infestations along roadsides and seedlings are of local importance in arable land.

Methods of control
S. ruwenzoriensis appears to be resistant to such growth-regulator herbicides as 2,4-D and MCPA. The shoots can be killed, but regrowth from the rhizomes occurs rapidly. Formulations containing dicamba, picloram or clopyralid are likely to be more effective.

S. moorei, on the other hand, lacks an underground rhizome system and is more susceptible. Young plants are readily controlled by 2,4-D and MCPA and established plants are severely checked. On older plants mixtures of 2,3,6-TBA and MCPA were found to give longer lasting control and dicamba mixtures are likely to be equally effective.

COMPOSITAE

Sonchus asper (L.) Hill   Spiny Sow-thistle
S. oleraceus L.   Sow-thistle

DESCRIPTION
These two annual species are erect herbs reaching a height of 1.2 m with stout, hollow, rather soft stems which exude a milky latex when cut. The leaves are spirally arranged, the flowers yellow and there is a long tap root. The leaves are more or less pinnately lobed, with an irregularly serrated margin and a pointed apex. The lower stem leaves have no distinct stalk and the basal pair of lobes project backwards past the stem. Flower heads are stalked and borne in a loosely branched, terminal inflorescence. They are 2–2.5 cm across and consist of numerous yellow, strap-shaped florets surrounded by several rows of overlapping bracts. The fruits are flattened and crowned with a ring of long, white, simple hairs. Unlike *Lactuca* and *Launaea* species, with which *Sonchus* species are easily confused, the apex of the fruit is more or less square instead of being gradually narrowed to a point.

The species are distinguished by the leaves. Those of *S. asper* usually have a prickly margin, the basal lobes are rounded and pressed against the stem, while the terminal lobe is usually narrower than the upper pair of lateral lobes. *S. oleraceus* leaves are generally more deeply divided, but not prickly. The terminal lobe is broader than the uppermost laterals and the basal lobes are pointed and spread away from the stem.

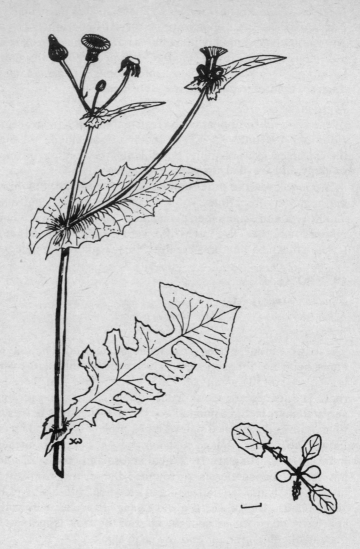

Figure 78. Sonchus oleraceus

Distribution and importance

Both species are native to Europe and North Africa, but have been introduced into many other areas and are now cosmopolitan. They occur widely in East Africa above about 1200 m and are reported as particularly common in the Machakos district of Kenya. *S. oleraceus* is generally rather commoner than *S. asper.*

These weeds are commonest in arable land and gardens in wetter areas. They produce large quantities of seed, can occur in large numbers and their luxuriant growth makes them very competitive with crop plants.

Methods of control

2,4-D, MCPA, 2,4-DB and MCPB at standard doses all give good control of *S. asper* and *S. oleraceus* seedlings, while pre-emergence treatment with 2,4-D is also reported to be successful. Mecoprop and dichlorprop are less effective. Ioxynil or bentazon give good control of young seedlings but resistance increases rapidly with age. Among residual chemicals atrazine, metribuzin, linuron and monolinuron are noted as effective pre- or early post-emergence treatments.

COMPOSITAE

Sphaeranthus suaveolens (Forsk.) DC.   Hardheads
S. bullatus Mattf.

DESCRIPTION

S. suaveolens is a branched, very leafy, annual herb, either growing along the ground with ascending branches, or erect to about 60 cm with broadly winged stems, alternate leaves, and purple, globular flower heads at the ends of the branches. The leaves are lanceolate, up to 10 cm long × 4 cm wide, pointed at the tip, with a toothed margin and no stalk, the blade continuing down into the wings of the stem. The flower heads are 1—2 cm in diameter and borne on erect, winged stalks up to 6 cm long. They consist of numerous crowded groups of tubular florets, each group surrounded by 3—8 narrow, pointed bracts. The head as a whole is also surrounded by several rows of bracts, of which only the tips are visible when the flowers are fully open.

S. bullatus is very similar in general appearance, but the leaves are smaller, up to 7 cm × 3 cm. The margins are curled backwards, and are more coarsely toothed, while the surface is puckered, with the veins very prominent below. The wings on the stem are also coarsely and irregularly toothed. The flower heads are smaller, about 8 mm diameter, while the bracts surrounding the groups of florets are more numerous.

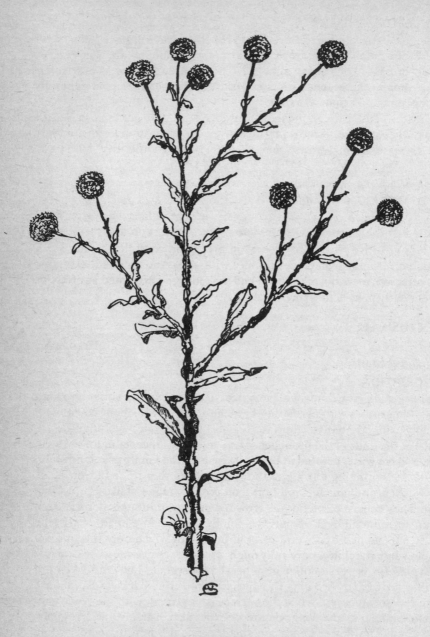

Figure 79. Sphaeranthus bullatus

Distribution and importance
Both species are indigenous. *S. suaveolens* is common and widespread throughout East Africa over a wide range of altitudes. *S. bullatus* is more restricted and occurs mostly at higher altitudes in the northern and central parts of Tanzania (it is very common in the Arusha district) and in the Kenya highlands. They grow as weeds in a wide range of situations, in arable crops, including coffee, particularly when irrigated, in fallow land (noted in the Naivasha district), in pastures and in lawns.

Methods of control
S. bullatus can be controlled in lawns with 2,4-D or MCPA but there is very little information on the susceptibility of *S. suaveolens* or on the effects of other herbicides. Another species in India is also reported as susceptible to 2,4-D.

COMPOSITAE

Spilanthes mauritiana (A. Rich.) DC.

DESCRIPTION
A creeping, perennial herb with opposite leaves and branched flowering stems up to 50 cm high terminating in solitary yellow flower heads. The stems are sparsely hairy and rooting at the lower nodes. The leaves are simple, up to 3 cm long × 2 cm wide, with stalks about 5 mm long and a smooth or shallowly toothed margin. The flower heads are on stalks up to 8 cm long. The bracts surrounding the heads are green and spreading or curved backwards. Within these and projecting beyond them are a ring of pale yellow ray florets and the disc consists of numerous yellow, tubular florets producing black seeds about 2 mm long. As the flowers age the receptacle bearing the florets elongates so that the head develops from a hemispherical to a conical shape.

Distribution and importance
Spilanthes is an indigenous species common at altitudes above about 1000 m in Kenya, Tanzania and Uganda. It is commonest as a weed of roadsides and waste ground and is reported to be a lawn weed of importance in Kenya.

Methods of control
No information is available on the control of this weed but, by analogy with other creeping perennial members of the family, it might be expected to show some resistance to 2,4-D or MCPA and greater susceptibility to dicamba formulations applied during a period of active growth.

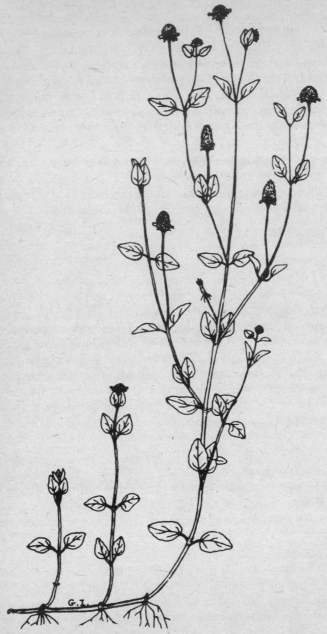

Figure 80. Spilanthes mauritiana

COMPOSITAE

Tagetes minuta L.

Mexican Marigold
Tall Khaki Weed

DESCRIPTION
The specific name of this weed refers to the size of the flowers rather than to the plant itself, which may be anything up to 2 m tall. It is an erectly branched annual herb with yellowish flowers, a furrowed stem and a strong aromatic scent. The leaves are opposite (though often alternate on smaller branches) up to about 25 cm long and pinnately divided into one terminal and several pairs of long, narrow lateral leaflets with sharply serrated margins. The flower heads are shortly stalked, up to 12 mm long, narrow and arranged, several together, in more or less flat-topped clusters at the ends of the branches. Each flower head consists of two yellow, tubular florets and two cream, strap-shaped florets, surrounded by 3 bracts which are united almost to the top and bear conspicuous orange-brown glandular patches. Fruits are black, spindle-shaped, covered with short, stiff hairs and bear 4 pointed scales at the apex, one larger than the others.

Distribution and importance
A species introduced from South or Central America, which is now common in the southern half of Africa, and in East Africa is one of the most widespread weeds. It is recorded from most agricultural districts and is one of the most important arable weeds at altitudes above about 1200 m. It is also common in waste land and often appears in great profusion after fires in upland areas or when upland forest is cleared and burnt. Seed is produced in very large quantities, and can germinate as soon as it is shed.

Methods of control
Seedlings and young plants are readily controlled by 2,4-D, MCPA, 2,4-DB, MCPB and other growth-regulator herbicides. Chemicals with a contact effect, such as bromoxynil, are effective against young seedlings and various residual herbicides belonging to the triazine and substituted urea groups can be used as pre- or early post-emergence treatments.

COMPOSITAE

Tridax procumbens L.

P.W.D. Weed

DESCRIPTION
A short-lived perennial, creeping herb branching at the base, with op-

Figure 81. Tagetes minuta

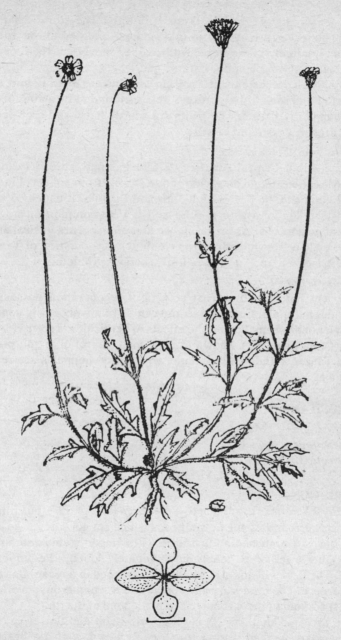

Figure 82. Tridax procumbens

posite, densely hairy leaves and cream flowers. The stem is hairy and brittle. The leaves are ovate, up to 5 cm long, pointed at the apex and narrowed gradually to the base, with coarse, deeply-cut teeth or lobes. In contrast to the procumbent stems and leaves, the flower heads are borne singly on erect stems up to 30 cm long and are up to 12 mm across. There are numerous, tubular disc florets surrounded by a ring of short, strap-shaped ray florets. The fruits are shortly hairy and crowned with long, stiff, straw-coloured bristles.

Distribution and importance
Introduced from Central America, *Tridax* has now spread throughout East Africa and is found on a wide range of soils from sea level in Zanzibar and Pemba to the Kenya highlands and Uganda. It usually occurs as a weed in fallow or waste ground and is very common on neglected lawns and in gardens. As an arable weed it often occurs in sisal and has been noted in cotton and other crops in the Mwanza district of Tanzania, though it does not appear to constitute a serious problem.

Methods of control
Tridax does not have the great powers of regeneration possessed by some other perennial Compositae and can be relatively easily controlled by cultivation. Information on the effects of herbicides is scanty, but 2,4-D and MCPA have been shown to give good control of young plants in sisal in Tanzania and similar results with these chemicals are reported from India.

COMPOSITAE

Volutaria lippii (L.) Maire
  (= *Centaurea lippii* L.)
V. muricata (L.) Maire
  (= *Centaurea muricata* L.)

DESCRIPTION
The genus *Volutaria* is similar to *Centaurea* (see p. 127) but differs in the bracts surrounding the flower heads which are not spiny.

  V. lippii is a branched annual herb with winged stems and alternate leaves, the lower ones deeply and pinnately lobed, the upper ones simple. The flower heads are purple and arranged in a loose, flat-topped inflorescence. The florets are all tubular and the numerous, overlapping bracts surrounding them come to a whitish point at the tip.

  V. muricata is similar in general growth habit and flower colour but the leaves are toothed, rather than deeply lobed, and the bracts surrounding the flower heads terminate in a whitish bristle.

Distribution and importance
Both species have been introduced from the Mediterranean region. *V. lippii* is well established in Kenya and is often found in dry, upland grassland. *V. muricata* has so far only been recorded from around Eldoret in Kenya.

Methods of control
There is no information concerning the effects of herbicides on these two species but various related *Centaurea* in Europe and Australia are reported to be susceptible to 2,4-D in the young rosette stage. Early spraying would be expected to give the best results.

CONVOLVULACEAE

Astripomoea hyoscyamoides (Vatke) Verdc.
   (= *Astrochlaena hyoscyamoides* (Vatke) Hall.f.)

DESCRIPTION
An erect, branched, short-lived perennial or annual herb, covered with greyish hairs, and reaching a height of 2 m with alternate, simple leaves and showy, white and purple flowers. The leaves are elliptic, up to 15 cm long × 6 cm wide, narrowed to a pointed tip and to a 2 cm stalk and have an entire or slightly indented margin. The flowers are shortly stalked and borne in small groups near the ends of the branches. In the upper part of the stem the leaves arise from the stalks of the flower clusters rather than from the stem itself. The corolla is funnel-shaped and 5-angled, 4 cm across and white with a purple throat. There are 5 pointed sepals, about 1 cm long, 5 stamens joined to the tube of the corolla and an ovary bearing a long style with 2 stigmas. The fruit is rounded and contains 4 triangular seeds.

Distribution and importance
A plant of drier grassland types and open bush at altitudes below 1200 m in Tanzania, Zanzibar and central and eastern parts of Kenya. It has not been found in Uganda. As a weed it is commonest in grassland, especially where there has been overgrazing, but it also occurs as an arable weed when grassland is newly opened up for cultivation. It was a common weed in the groundnut areas near Kongwa in Tanzania. In grassland it is unpalatable to stock.

Methods of control
Trials in Tanzania have shown that *Astripomoea* is very sensitive to 2,4-D and MCPA and young plants are controlled by low doses. It can, therefore, be eradicated from grassland without difficulty and control in maize is also possible. The effects of other chemicals are not known.

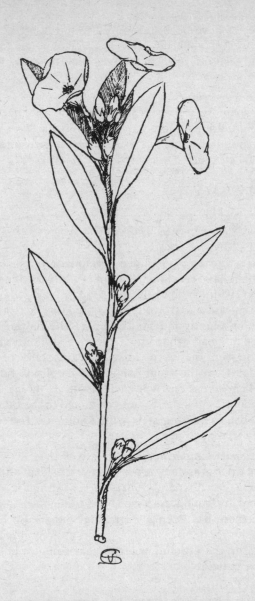

Figure 83. Astripomoea hyoscyamoides

CONVOLVULACEAE

Cuscuta kilimanjari Oliv.   Dodder

DESCRIPTION

An annual, parasitic herb with slender, yellow, leafless stems which twine round the stems of suitable host plants and bear cream coloured flowers. The stems are attached to the host by means of suckers. The leaves are reduced to small scales and the flowers are borne on short stalks in loose, lateral clusters. The flowers are 4 —9 mm across and consist of a calyx of 5 blunt sepals and a tubular corolla with 5 blunt lobes, attached to which are 5 short stamens and thin, fringed scales. The ovary bears 2 short styles and develops into a capsule which splits round the base to release the blackish seeds.

Distribution and importance

This native dodder occurs widely throughout East Africa between 500 and 2600 m, growing mostly on a variety of native plants on the fringes of forested areas. It can also parasitise coffee and is noted as becoming a problem in the Kiambu district.

Methods of control

There is very little information on *C. kilimanjari* but dodder species generally produce much dormant seed, which can remain viable for long periods and makes eradication very difficult. In Australia a recommended treatment involves burning or spraying the host with diquat and treating with a soil-sterilant to prevent re-establishment and this drastic treatment may be worth considering if it is desired to prevent the spread of a small, initial infestation. Low rates of glyphosate have been used to kill established dodder selectively in lucerne and propyzamide appears to be one of the more effective residual herbicides against *Cuscuta* species. Pendimethalin has also been found to prevent the establishment of *Cuscuta* and a combination of this chemical with linuron and diuron has been reported an effective treatment in Europe.

CONVOLVULACEAE

Dichondra repens J. R. & G. Forst.   Kidney Weed

DESCRIPTION

A perennial herb, creeping along the ground, with slender stems up to 60 cm long which root at the nodes. The simple leaves are arranged alternately. They have a rounded, kidney-shaped blade up to 4 cm across borne on an upright stalk up to 5 cm long. When young the leaves are covered with silky hairs. The flowers are small and inconspicuous, white

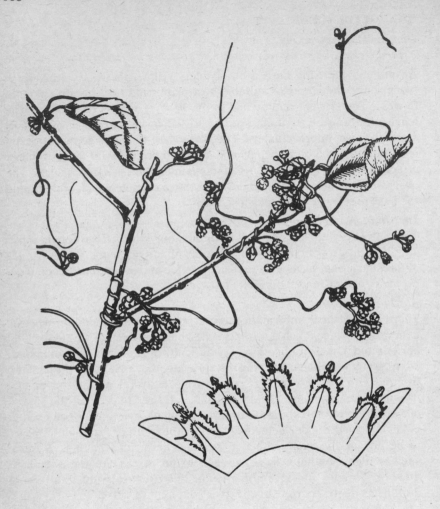

Figure 84. Cuscuta kilimanjari

or greenish in colour and borne singly on stems 5 mm or less long in the leaf axils. There are 5 sepals about 3 mm long, and a slightly shorter corolla of 5 petals, joined for about 2/3 of their length, with 5 stamens attached. The ovary is deeply 2-lobed and bears 2 styles.

Distribution and importance

Dichondra is distributed widely through the tropics and in East Africa is of common occurrence at altitudes between 900 and 2500 m. It is found both in grassland and cultivated areas, but is most troublesome as a weed of lawns where, because of its closely creeping habit, it is not affected by mowing. In shaded parts of lawns it can sometimes virtually replace the grass.

Methods of control

In Australia both MCPA and 2,4-D are reported to control this weed at normal rates for lawn use, but two applications are often needed to obtain a complete kill and formulations containing dicamba have been found more effective in New Zealand. As with all lawn weeds, the herbicide treatment is best applied during a period of rapid growth.

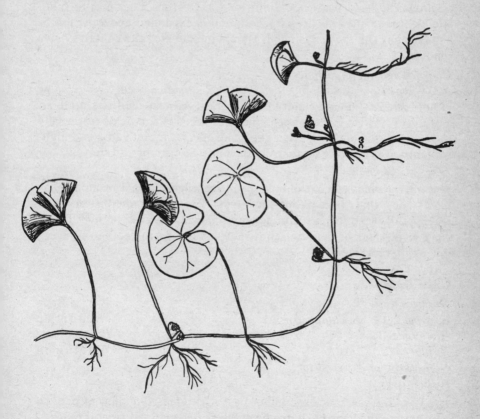

Figure 85. Dichondra repens

CONVOLVULACEAE

Ipomoea aquatica Forsk.   Swamp Morning-glory

DESCRIPTION

I. aquatica is a trailing annual or, under suitable conditions, a stoloni-
ferous perennial herb of wet places with alternate, triangular-shaped
leaves and large funnel-shaped, solitary flowers, purple, pink or white in
colour. The stems may be up to 3 m long and root at the nodes. The leaves
have a stalk which may be up to 25 cm long and a blade up to 15 cm long
× 9 cm wide, which may be pointed or rounded at the angles. The
flowers arise in the leaf axils, also on long stalks, and are usually about 5
cm long. The 5 sepals are 6 —12 mm long, the corolla has a narrow tube,
to which the 5 stamens are attached, and is expanded above with a 5-
pointed rim. The capsule is about 10 mm in diameter and contains hairy
seeds.

Distribution and importance

A species found throughout the tropics, *I. aquatica* occurs on swampy
ground and in stagnant water in all the East African countries up to an
altitude of 1200 m. It is not uncommon as a weed of rice fields, especially
in Tanzania. In South-East Asia the same species is cultivated and highly regarded as a leaf vegetable and flavouring for soups and stews.

Methods of control

Few observations on control have been recorded but in Tanzania it has
been noted that, because of the presence of submerged stems and
leaves, this weed increases in density following the use of such herbicides as propanil or bentazon in rice. Some control appears to be possible with oxadiazon.

CRUCIFERAE

Brassica napus L.   Rape
B. rapa. L. (= *B. campestris* L.)   Wild Turnip
B. oleracea L.   Wild Cabbage
B. juncea (L.) Czern.   Wild Mustard
B. integrifolia (West) Rupr.

DESCRIPTION

The five *Brassica* species found in East Africa are all similar in general
appearance and have been much confused. They are all erect, branched
annuals or biennials up to 1 m or more high with large, alternate,
variously lobed leaves, racemes of yellow flowers and long narrow fruits.
    In *B. napus* the upper stem leaves have no stalk and the broad basal

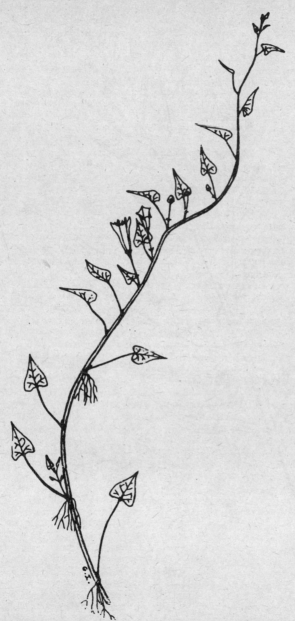

Figure 86. Ipomoea aquatica

Figure 87. Brassica napus

lobes clasp the stem. The lower leaves have a distinct stalk, a few coarse hairs, a large coarsely toothed terminal lobe and several pairs of smaller lobes divided to the midrib. They are bluish green in colour. The flowers are pale yellow and borne on slender stalks, in flat-topped clusters, at the ends of the branches. As it develops, the inflorescence lengthens so that the unopened flower buds overtop the opened flowers. The flowers are about 2 cm across, with 4 petals about twice as long as the sepals and an ovary which develops into an upright fruit up to 10 cm long and containing black spherical seeds.

B. rapa differs most obviously from the above in that the unopened flower buds remain below the level of the opened flowers, in the lower leaves being a brighter green and bristly and in the brighter yellow flowers. In *B. oleracea* the lower leaves are fleshy and hairless, with prominent whitish nerves, and the flowers are larger, 3 —4 cm across, in a much elongated inflorescence.

The other two species differ in having the upper stem leaves on distinct stalks, not clasping the stem. In *B. juncea* the lower leaves have a large terminal lobe and up to 3 pairs of small, lateral lobes. The tip of the fruit forms a tapered, seedless beak 6 —12 mm long. *B. integrifolia* has the lower leaves either entire or with a single pair of small, lateral lobes and the beak at the tip of the fruit is shorter (2 —7 mm).

Distribution and importance

These *Brassica* species are widely grown as crops and have become naturalised as weeds of cultivation in many parts of the world. *B. napus* occurs mainly in Kenya at altitudes of 1750 —2300 m and has become an important weed of cereals and other crops in Uashin Gishu, Laikipia and Naivasha. It has sometimes been referred to as 'charlock', a name properly referring to *Sinapis arvensis,* a species of doubtful occurrence in East Africa.

B. rapa and *B. integrifolia* are distributed more widely in Kenya, Tanzania and Uganda, the former recorded between 1500 and 2600 m, the latter between 0 and 2000 m (and probably indigenous in Ethiopia). *B. oleracea* and *B. juncea* have only been noted in Kenya and Tanzania, the former from relatively high altitudes, 800 —2000 m, and only occasionally as a weed, the latter from 50 —2000 m.

Methods of control

Like many annual weeds of the Crucifer family the *Brassica* species are susceptible to 2,4-D, MCPA and other growth-regulator type herbicides. In the young stages they are also susceptible to various contact herbicides and germinating seedlings can be controlled with a range of residual chemicals including triazines, substituted ureas, EPTC and pendimethalin.

CRUCIFERAE

Capsella bursa-pastoris (L.) Medic. Shepherd's Purse

DESCRIPTION

An annual herb with a tap root and a basal rosette of leaves, from the centre of which arise one or more stems, up to 50 cm high, bearing loose cylindrical heads of small, white flowers. The basal leaves reach a length of 15 cm and are oblanceolate, with a blunt tip and the base narrowed gradually into the stalk. They may be either entire or deeply and pinnately lobed. The stem leaves are smaller, alternate and unstalked, and have auricles projecting at the base and clasping the stem. The flowers are numerous and are borne on slender stalks up to 2 cm long on the upper part of the stems. They are 3 mm across and consist of 4 sepals, 4 petals about twice as long as the sepals and 6 stamens. The ovary has a short style and develops into a flattened fruit of a characteristic, rounded-triangular shape with an indented apex.

Distribution and importance

A common weed of arable land distributed mainly through the temperate parts of the world, but also introduced in various tropical countries. In East Africa it occurs at altitudes around 1500—2000 m and is commonest in cereal crops in the Kenya highlands, especially in the Naivasha district. It has also been recorded in the Kigezi district of Uganda and in northern Tanzania, the Usambaras and the Southern Highlands. Although readily controlled by cultivation, *Capsella* produces large quantities of seed and dense infestations can form quite a serious weed problem.

Methods of control

Capsella can be controlled with MCPA, 2,4-D and other growth-regulator type chemicals. Contact herbicides including diquat and paraquat are only effective when applied to young seedlings. Among residual chemicals the substituted ureas including linuron and metoxuron and triazines, including atrazine and metribozin, give good pre- and early post-emergence control.

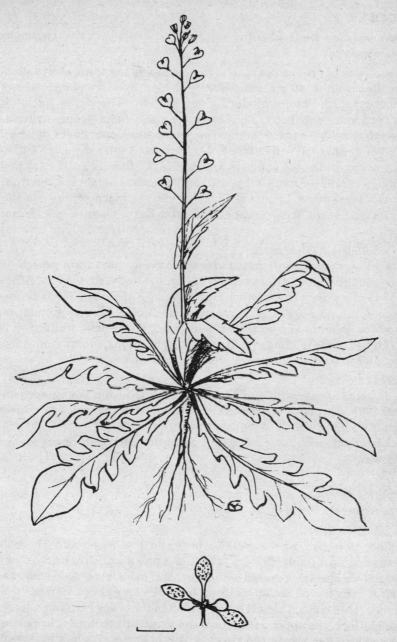

Figure 88. Capsella bursa-pastoris

CRUCIFERAE

Coronopus didymus (L.) Smith   Twincress

DESCRIPTION

An annual or short-lived perennial herb with a taproot, numerous prostrate stems up to 40 cm long, finely divided leaves and racemes of
minute greenish flowers producing paired fruits. The stems are much
branched and finely hairy. The leaves are stalked and deeply and pinnately divided into narrow, pointed segments, the upper edges of which
are often further toothed or lobed. The numerous inflorescences are up
to 5 cm long and borne opposite the leaves. The flowers are 1—1.5 mm
in diameter on 2—3 mm stalks, with inconspicuous petals and the fruit,
up to 1.7 mm long × 2.5 mm wide, splits into two, each section containing a single seed. When crushed the plant has a distinct, unpleasant
smell.

Distribution and importance

C. didymus probably originated in South America but is now present in
many countries around the world. It has been introduced into East Africa
and has been recorded from various parts of Kenya and Tanzania between 1350 and 2800 m. It is mainly a weed of disturbed ground, on
roadsides, in crops and occasionally in grassland. In dairy pastures it is
reported to cause tainting of milk. Although not of major importance it is
a persistent weed and likely to spread.

Methods of control

C. didymus is susceptible to MCPA and 2,4-D in the young seedling
stage but older plants are relatively resistant. Older plants also recover
readily from the effects of diquat or paraquat. Pre- or early post-emergence treatment with substituted urea or triazine chemicals appears to be
effective.

CRUCIFERAE

Erucastrum arabicum Fisch. & Mey.

DESCRIPTION

A much branched, annual herb, growing to 1 m high, with alternate
leaves and small, white flowers (sometimes pale purplish or pale yellow).
The stem is sparsely covered with short, stiff hairs lying close along the
surface. The leaves are up to 15 cm long; the lower ones stalked, with
pinnate lobes divided almost to the midrib in the lower part and a larger
terminal lobe, the upper ones sessile and irregularly toothed. The flowers
are about 10 mm across and borne on 5 mm long stalks in a branched,
terminal inflorescence. They consist of 4 sepals, 3 mm long, 4 somewhat

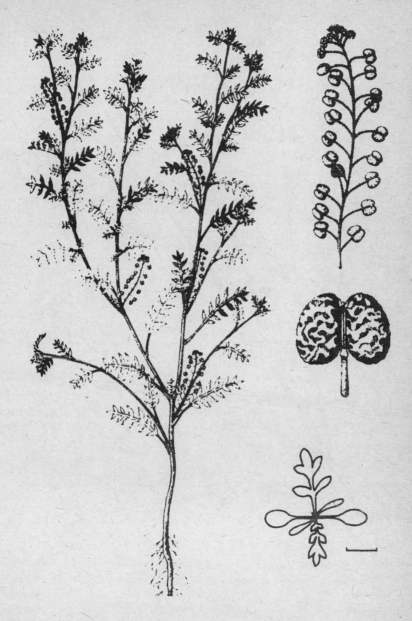

Figure 89. Coronopus didymus

Figure 90. *Erucastrum arabicum*

longer petals, 6 stamens and a 2-celled ovary. The fruit is narrow, up to 4 cm long, 4-angled in cross section and with a short, pointed, terminal beak. The seeds are round, about 2 mm across and dark brown in colour.

Distribution and importance
An indigenous species distributed throughout East Africa, mostly at medium altitudes, between 900 and 2000 m, and usually occurring as an arable weed. In Kenya it is reported mostly from the highland areas, in Tanzania from the northern, Usambara and Kongwa districts and, in Uganda, from the neighbourhoods of Kampala, Mbarara and Kigezi. It occurs in a wide range of arable crops, especially in cereals, and is common in gardens.

Methods of control
There is little specific information on the reaction of *Erucastrum* to herbicides, but it appears to be very susceptible to 2,4-D, MCPA and other growth-regulator type chemicals, as would be expected from its close relationship to *Brassica* species.

CRUCIFERAE

Raphanus raphanistrum L.   Wild Radish, White Charlock

DESCRIPTION
An annual herb up to 1 m high, similar in general appearance to *Brassica napus,* but with larger, usually white or purple flowers. Yellow-flowered forms also occur, but are not common. The stem and leaves are bristly, with stiff hairs. The lower leaves are stalked, with a large, coarsely toothed terminal lobe and several pairs of smaller lateral lobes. The upper leaves are smaller and less deeply divided. The flowers are arranged in elongated, terminal racemes, and are 1—2 cm across. The 4 petals are twice as long as the sepals and there are 6 stamens. The fruit is elongated, with a pointed tip, and is constricted between the 3—8 seeds. When ripe it readily breaks up at the constrictions into 1-seeded joints.

Distribution and importance
Another cereal weed introduced from temperate regions, *Raphanus* is established in parts of the Kenya highlands at altitudes around 1800 m, but has not so far been recorded in Uganda or Tanzania. It is commonest in the Uasin Gishu, Laikipia, Naivasha and Molo districts, mostly in cereals, but also in other arable crops, and appears to be spreading. The 1-seeded segments of fruit are similar in size to small-grain cereals and are difficult to eliminate from seed grain.

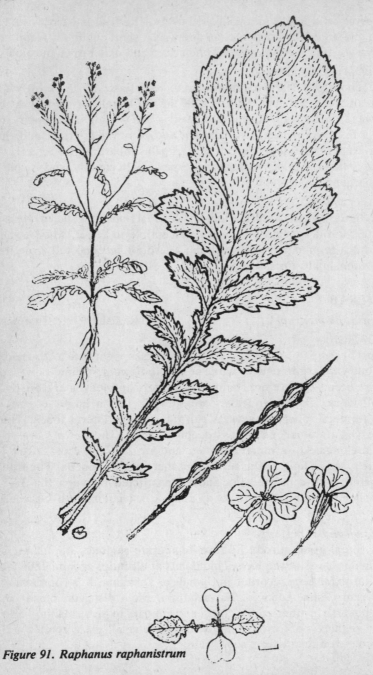

Figure 91. Raphanus raphanistrum

Methods of control

Like the *Brassica* species, *Raphanus* is readily controlled by 2,4-D, MCPA and mecoprop (less readily by 2,4-DB and MCPB) and in the young stages by contact chemicals. Pre-emergence control is also possible with several types of residual chemicals, including substituted ureas and triazines but, like *Capsella*, it has a fast-growing taproot and established seedlings are much more resistant.

EUPHORBIACEAE

Euphorbia hirta L.   Asthma Weed
E. prostrata Ait.   Blueweed
E. inaequilatera Sond. (= *E. sanguinea* Steud. & Boiss.)

DESCRIPTION

All three are small, prostrate, annual herbs, with milky sap, opposite leaves and inconspicuous greenish flowers. The plants often have a reddish tinge.

All three are small, prostrate, annual herbs, with milky sap, opposite leaves and inconspicuous greenish flowers. The plants often have a reddish tinge.

E. hirta has a tap root, from which develop several little-branched stems, creeping along the ground at first, later becoming erect. The leaves are simple in shape, up to 4 cm long × 1 cm wide, with a short stalk and finely serrated margin. At the base, the side away from the stem is rounded, the other narrowed gradually to the stalk. The flowers are minute and form dense, rounded, almost sessile clusters in the axil of the leaves. The fruits are small capsules containing 3 seeds.

E. prostrata is similar, but smaller, with numerous slender branches up to about 10 cm long, radiating from the crown. They are prostrate and have a line of hairs on the upper side. The leaves are bluish green in colour, about 5 mm long and rounded both at apex and base. The minute flowers are borne in the leaf axils, mostly in pairs.

E. inaequilatera is characterized by very asymmetrical leaves up to 12 mm long × 6 mm wide which are distinctly toothed on the side away from the stem. The flowers are borne on short, leafy branchlets, one flower to each pair of minute leaves, or sometimes in clusters of 2 or 3 together, never in dense clusters as in *E. hirta*.

Distribution and importance

E. prostrata appears to have been introduced from tropical America, the others are native species. All are widespread through East Africa, *E. hirta* being the commonest. They occur both in arable land and poor

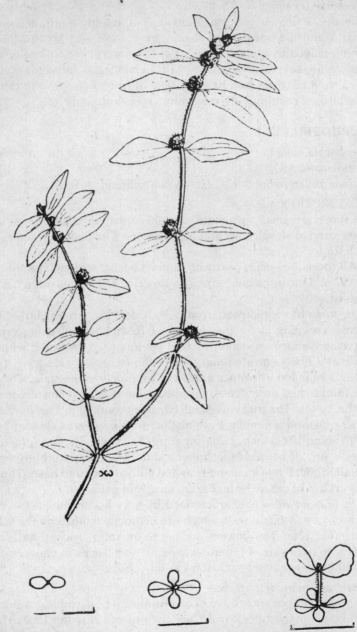

Figure 92. Euphorbia hirta

grassland. Abundant seed is produced which germinates rapidly under moist conditions and plants produce seed at a very early stage so that dense infestations build up rapidly. They are particularly common in gardens and lawns, where their prostrate habit is well adapted to the survival of mowing.

Methods of control
Information on susceptibility to 2,4-D is somewhat conflicting. *E. hirta* and *E. prostrata* are reported to be resistant in Kenya, but in Australia they are regarded as very susceptible. 2,4-D and MCPA have also given good control of *E. hirta* in Zimbabwe and in sisal nurseries in Tanzania, and MCPB has been effective in Trinidad. Other *Euphorbia* species generally appear to be relatively susceptible to the effects of growth-regulator herbicides, and to pre-emergence treatment with trifluralin or diuron.

EUPHORBIACEAE

Ricinus communis L.   Castor Oil Plant
DESCRIPTION
A large, bushy plant reaching a height of about 2 m when grown as an annual crop, but developing into a small tree if left for a few years. It has large, alternate, leaves and spikes of small cream or reddish flowers. The leaves are 30 cm or more in length and breadth and are palmately divided into 5—9 lobes, with pointed tips and serrated margins. They are borne on stalks 30 cm or more long, with conspicuous glands. The flower spikes are borne in the axils of the upper leaves and are 15—30 cm long, with female flowers in the upper part of the spike and male flowers below. Neither type of flower has petals, the male has numerous, cream-coloured stamens and the female 3 red stigmas. The fruits are spherical, green at first, but later becoming brown, covered with soft spines and 1—2 cm across. When ripe they split to release 3 large, mottled see s.

Distribution and importance
Much grown as a crop throughout the tropics and widely distributed, both as a crop and a weed, in Kenya, Uganda, Tanzania and Zanzibar, from sea level to 2500 m. It is principally a weed of waste ground and its importance lies mainly in the fact that the seed coat is poisonous to animals and poultry.

Methods of control
Hoeing seedlings or cutting mature plants is all that is usually required, as *Ricinus* rarely develops into dense infestations. Where such mea-

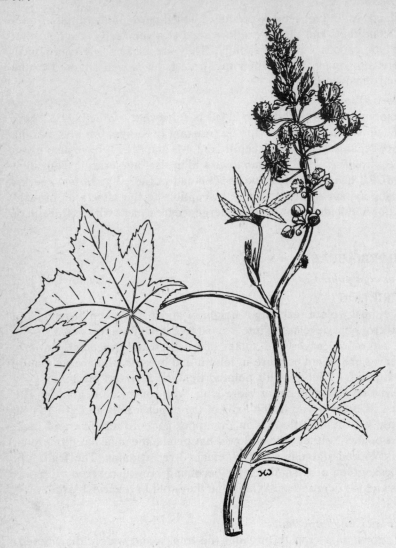

Figure 93. Ricinus communis

sures are insufficient overall foliage application of 2,4-D is effective against smaller plants and glyphosate also gives a good kill. In the USA picloram has been found effective applied in pellet form as a soil treatment.

GERANIACEAE

Geranium arabicum Forsk.　　　　　　　　　　　Cranesbill
(= *G. simense* A. Rich.)

DESCRIPTION

A scrambling, runner-producing, perennial herb, with stems up to 1 m or more long, alternate, palmately divided leaves and paired, pink or white flowers. The leaves are hairy, the stalks up to 15 cm long, the blades more or less circular in outline, up to 15 cm in diameter and divided ¾ of the way across into 5 segments, the segments themselves pinnately divided into lobes with pointed tips. The flowers are borne on inflorescence stalks up to 10 cm long, arising in the leaf axils, and the individual flowers of each pair have stalks up to 4 cm long. The flowers are 2 cm across and consist of 5 pointed sepals, 5 red-veined petals about the same length and indented at the tip, 10 stamens and a 5-lobed ovary extended into a beak at the tip. Each lobe of the fruit contains a single seed.

Distribution and importance

A plant of damp, shady habitats of mountainous regions in many parts of Africa, *G. arabicum* occurs mostly at altitudes above 1800 – 2000 m in East Africa. It is common in the Kenya highlands, has been reported in the northern part of Tanzania and, although not recorded as a weed in Uganda, seems certain to occur in suitable habitats. It is principally a weed of arable land and infests most high-altitude crops, especially cereals and pyrethrum. *G. aculeolatum* Oliv. with prickly stems and *G. ocellatum* Cambess. an annual species, are also recorded as weeds.

Methods of control

The scanty information available on this species suggests that, in cereals, it can only be partially controlled by 2,4-D and MCPA. In Britain several other annual and perennial *Geranium* species have been found moderately resistant to these two chemicals but formulations containing dicamba, clopyralid or picloram offer better chances of success.

LABIATAE

Leonotis nepetifolia (L.) Ait.f.　　　　　Lion's Ear, Lion's Tail
L. mollissima Guerke

DESCRIPTION

L. nepetifolia is an erect, branched annual herb growing to 1.5 m high, with a stout, 4-angled stem, paired, simple leaves and dense whorls of orange flowers. The lower leaves are up to 8 – 10 cm long, with a stalk

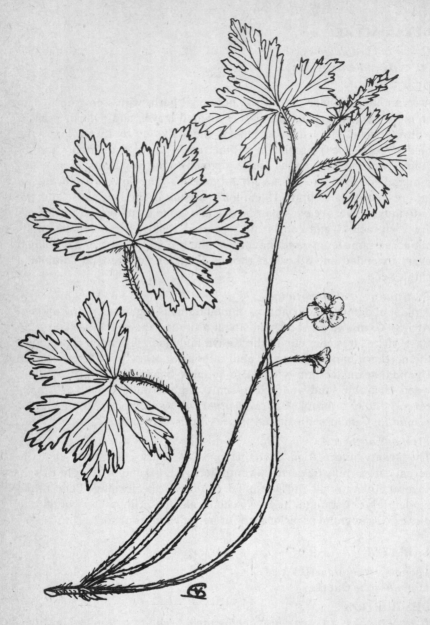

Figure 94. Geranium arabicum

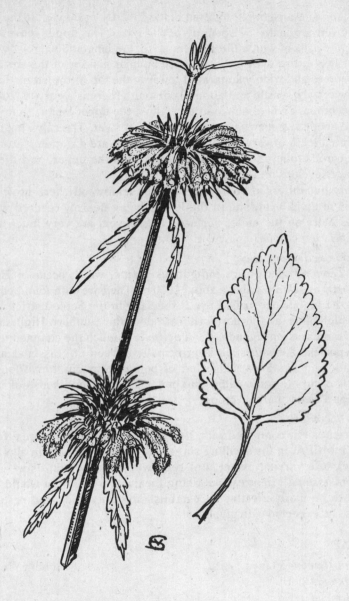

Figure 95. Leonotis mollissima

2 —3 cm long and a coarsely toothed blade, bluntly narrowed at the tip and narrowed gradually or abruptly at the base. The upper leaves or bracts, in the axils of which the flowers arise, are long and narrow, with a more or less entire margin. The dense circular masses of flowers are 5 —6 cm across and widely separated towards the top of the stem. Individual flowers are sessile and about 2 cm long. There is a curved, tubular, hairy corolla, divided above into two lips, the upper longer than the lower and hooded, somewhat resembling a lion's ear. The calyx bears 5 long, pointed teeth of very unequal length. There are 4 stamens attached to the corolla tube, the lower pair longer than the upper, and the 4-lobed ovary develops into a fruit of 4 nutlets.

L. mollissima differs in being a perennial, semi-woody herb growing to a height of up to 3 cm, and in the leaves being densely covered with soft white hairs on the under surface. The flowers are very much the same as in *L. nepetifolia*.

Distribution and importance

The two *Leonotis* species are indigenous plants, which occur in East Africa mostly at altitudes above about 1000 m. They are both found commonly in the higher parts of Kenya, especially in the Sotik district and the Kisii highlands, and in northern Tanzania, the Southern Highlands and the Usambaras. In Uganda, *L. nepetifolia* is much the commoner of the two species. *L. mollissima* is principally a weed of grassland and waste land, and has been suspected of being poisonous to cattle. *L. nepetifolia* occurs to some extent in similar habitats, but is also of importance as a weed of arable crops.

Methods of control

Both species can be controlled with 2,4-D, or perhaps rather more effectively with MCPA, in the seedling stage. Established plants can also be killed back to a varying extent, but regrowth often occurs, especially with *L. mollissima*. Mixtures containing dicamba or picloram would be expected to be more effective on established plants and good results would also be expected with glyphosate.

LABIATAE

Leucas martinicensis (Jacq.) Ait.f.   Bobbin Weed
L. neuflizeana Courb.

DESCRIPTION

L. martinicensis is an erect, softly hairy, sparingly branched annual herb, growing 0.6 —1 m high, with paired leaves and white flowers in dense, more or less globular clusters round the stem. The stems are

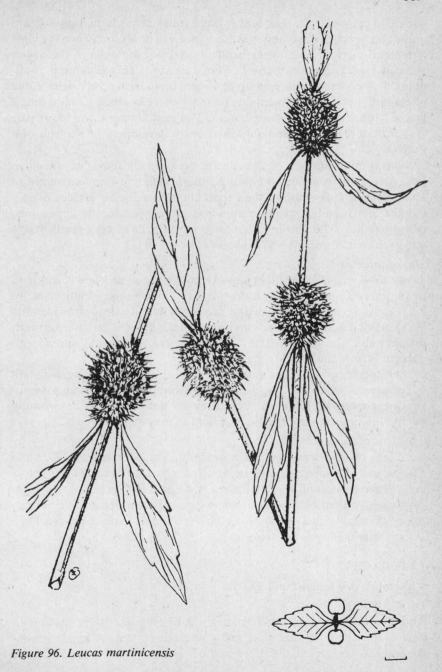

Figure 96. Leucas martinicensis

square in cross section. The leaves have stalks about 1 cm long and are entire, up to about 8 cm long × 2.5 cm broad, with a bluntly and regularly toothed margin. The flower heads arise from the axils of the upper pairs of leaves, contain many flowers and are widely separated. Individual flowers are sessile and up to 15 mm long, with a very hairy calyx of 5 sharply pointed segments, a tubular, curved corolla divided into 2 lips of approximately the same length, 2 pairs of stamens, the lower pair longer than the upper, and a 4-lobed ovary developing into a fruit of 4 nutlets.

Several other species of *Leucas* are occasionally found as weeds in East Africa, the commonest being *L. neuflizeana* growing to a height of about 30 cm. It is distinguished from the above by its sessile, deeply toothed leaves and by its more numerous flower heads, the upper ones close together. The flower heads arise in the axils of very small bracts and each contains only 8 —12 flowers.

Distribution and importance
L. martinicensis is widely distributed through the tropics and is common in the three East African countries over a wide range of altitudes. In Kenya it is recorded as a weed in the Molo and Kitale districts and is very common in the Machakos area. In Tanzania it is noted from northern districts and the Ilonga region. *L. neuflizeana* is also widespread and especially common near Machakos.

Both species occur either in grass or arable land, often appearing late in the season when the crop has become established. Although sometimes occurring in large numbers they do not appear to be strongly competitive species and are not regarded as very serious weeds.

Methods of control
Seedlings of *L. martinicensis* are moderately susceptible to 2,4-D and MCPA and a related species in Mauritius is also susceptible to pre-emergence treatment. A related species in the Sudan is reported as susceptible to pre-emergence treatment with oxadiazon. No specific information is available on *L. neuflizeana,* but it would be expected to react in a very similar way to *L. martinicensis.*

LYTHRACEAE

Ammannia prieuriana Guill. & Perr
DESCRIPTION
An erect, little-branched, glabrous, annual herb, 30 —60 cm high, with narrow leaves in opposite pairs and numerous clusters of small, greenish flowers. The stem is sharply 4-angled. The leaves are stalkless, up to

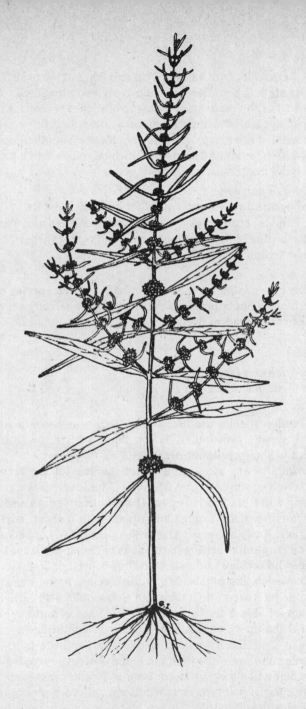

Figure 97. Ammannia prieuriana

10 cm long × 1 cm wide, with a pointed tip and 2 small lobes at the base which clasp the stem. The inflorescence is leafy and elongated, occupying the upper half of the main stem and side branches. It consists of numerous, shortly separated whorls made up of branched flower clusters arising in the axils of the leaves. The flowers have a 5-toothed calyx, inconspicuous whitish petals and an ovary which develops into a brownish capsule containing numerous, very small seeds.

Distribution and importance

A. prieuriana occurs in swamps and other aquatic habitats in various parts of Africa and is recorded as a weed of rice in Kenya and Tanzania. It is reported to be one of the more numerous and competitive rice weeds on the Mwea Irrigation Scheme in Kenya.

Methods of control

No information is available on the susceptibility of this species of *Ammannia* to herbicides but bentazon is reported to give effective control of a related American species in rice.

MALVACEAE

Abutilon mauritianum (Jacq.) Medic.
A. guineense (Schumach.) Bak. & Exell

DESCRIPTION

These two *Abutilon* species are much branched, tough-stemmed herbs or softwooded shrubs growing to 1.5 m high, with large, alternate, simple leaves and showy yellow flowers.

In *A. mauritianum* the stems and under surfaces of the leaves are covered with a velvety layer of greyish hairs. The leaves have stalks 15 cm or more long and a blade up to about 15 cm both in length and breadth, gradually narrowed at the apex and tipped with a short, stiff point. The leaf margin is finely serrated. The flowers are 2.5 —4 cm across and borne singly on long stalks arising in the axils of the upper leaves on the main branches and on short side shoots. They consist of 5 sepals joined at the base, 5 overlapping petals with reddish veins, numerous stamens with their filaments joined into a tube and a ring of 25 —40 carpels. The carpels develop into black fruits containing 2 —3 seeds and have a long straight point at the tip which is 1/3 the total length of the fruit.

A. guineense is similar in general appearance, but the hairs on the stem and leaves are more white than grey, and the leaves are smaller and wrinkled, the blade being up to 10 cm long and rather less wide, while the apex is not sharply pointed. The carpels are only very shortly pointed and are covered with soft hairs.

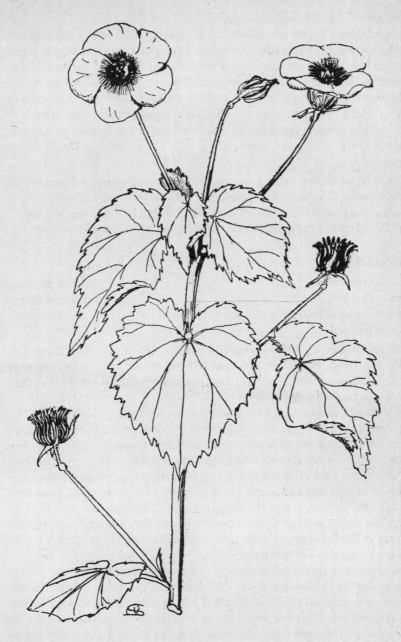

Figure 98. *Abutilon mauritianum*

Distribution and importance
A. *mauritianum* is the commoner of the two species and is distributed over a wide range of altitudes throughout Kenya, Tanzania, Uganda and Zanzibar. A. *guineense* is reported more frequently from Tanzania than elsewhere. Both species are indigenous plants of grassland or forest margins and are commonest as weeds in pastures and waste ground in areas of higher altitude and rainfall. They are of general importance in reducing the grazing potential of grassland and often appear as a result of overgrazing.

Methods of control
Abutilon species are among the easier grassland weeds to control by cutting, as their capacity for regrowth is relatively limited. Young shoots can also be killed back with 2,4-D or MCPA, but, for a complete kill of established plants, formulations containing picloram are more effective.

MALVACEAE

Hibiscus trionum L.   Flower-of-an-hour
H. cannabinus L.   Kenaf, Deccan Hemp
H. mastersianus Hiern

DESCRIPTION
There are numerous *Hibiscus* species in East Africa, of which those listed are the most important as weeds. All three are tall, erect, coarsely hairy, tough-stemmed, annual herbs with alternate, palmately veined leaves and large, showy flowers.

H. trionum grows to 1.2 m. The leaves have stalks about 5 cm long and blades up to 6 cm across, palmately divided almost to the base into 3—5 lobes, which are themselves irregularly and pinnately lobed. The flowers are 3—4 cm across, white, cream or yellow with purple centres and are borne on 5 cm stalks in the axils of the upper leaves. They are made up of 5 petals and a calyx of 5 broadly triangular lobes joined almost to the tip with a surrounding ring of 12 long, narrow bracts. The numerous stamens have their filaments joined into a tube. The fruit is a capsule about 1 cm across and is enclosed by the dry, papery calyx which becomes enlarged as the fruit ripens.

H. cannabinus is taller, reaching 2 cm or more, and has tough, prickly stems. The leaf stalks are up to 25 cm long, the blades 15 cm across, deeply and palmately divided into 3—7 narrow, toothed lobes. The flowers are up to 10 cm across, pale yellow, white or greyish with a purple centre and are borne on very short stalks in the leaf axils. The bracts surrounding the calyx are 7—8 in number. The calyx teeth are long and

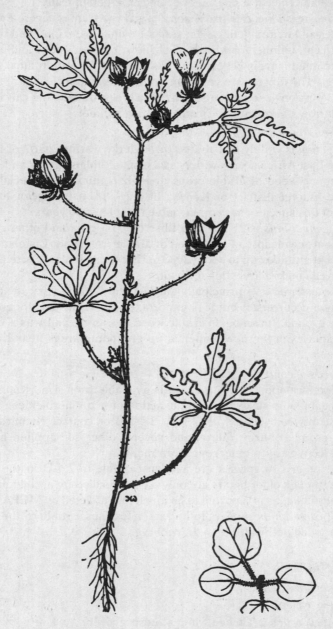

Figure 99. Hibiscus trionum

pointed and the calyx does not become enlarged in fruit.

H. mastersianus can also reach 2 m and the stem is covered with short prickles and irritant hairs. Leaf stalks are up to 8 cm long and the blades 10 −15 cm, unlobed or with 3 −5 shallow, palmate lobes and an irregularly toothed margin. The midrib has a distinctive longitudinal split near its base. The flowers are yellow or orange with a purple centre and up to 6 cm across on 4 −8 cm long stalks. The bracts outside the calyx are long and narrow, forked at the tip and 9 −10 in number.

Distribution and importance

H. trionum is widely distributed through the warmer parts of the world and in East Africa occurs widely as a plant of upland grassland. It is also a common weed of arable crops at higher altitudes, especially in the Kitale-Eldoret district of Kenya. In the USA it has been found that seed of this species can remain viable for at least 20 years.

H. cannabinus is grown as a fibre crop in India and other countries, and is also widespread as a weed of arable crops and wasteland in East Africa at altitudes up to about 3000 m. In cereals its large size and tough stems can make harvesting difficult.

H. masterianus is principally an East and Central African species, of common occurrence, but less important as a weed, usually growing on waste ground. It can be an arable weed, however, and was a very common species in the newly opened up groundnut areas near Kongwa in Tanzania.

Methods of control

Much information on *H. trionum* is available from American work and it appears to be susceptible to the action of many herbicides. Seedlings are readily controlled with 2,4-D, 2,4-DB or contact chemicals, while atrazine, metribuzin, diuron and several other soil-applied herbicides are effective as pre-emergence treatments.

The other two species are also susceptible to 2,4-D in the seedling stage, though older plants are only partially killed by normal doses, and *H. mastersianus* is reported to be much less affected by MCPA than 2,4-D. *H. cannabinus* appears to be rather less susceptible to the action of soil-applied herbicides than *H. trionum*.

MALVACEAE

Malva verticillata L.   Mallow

DESCRIPTION

An erect, much branched, annual herb growing to 1 m with rounded, alternate leaves and pale pink flowers. The leaves have a long stalk and

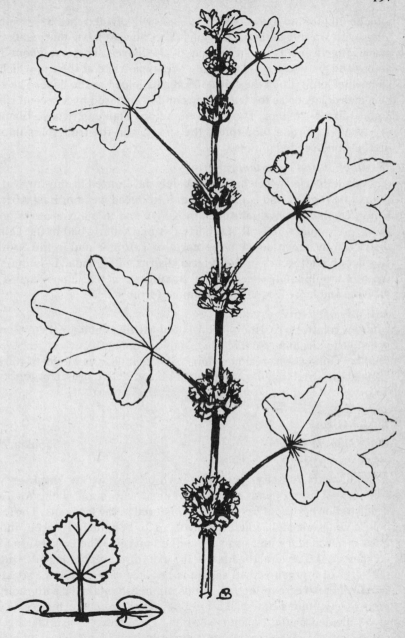

Figure 100. Malva verticillata

a blade up to 8 cm across which is shallowly divided into 5—7 palmate lobes with sharply serrated margins. The flowers have short stalks and are arranged in clusters in the axil of the leaves. They are about 1 cm across and are made up of 5 petals, deeply notched at the tip, which are somewhat longer than the calyx of 5 sepals joined at the base. The calyx becomes enlarged as the fruit ripens and is enclosed by a second ring of 3 sepal-like segments. There are numerous stamens, whose filaments are joined to form a tube round the styles, and the fruit splits into numerous, one-seeded nutlets.

Distribution and importance
An Asiatic species which has been widely naturalized in sub-tropical and temperate regions and is probably an introduced weed in East Africa. It is now common at high altitudes in Kenya and in many areas is one of the worst annual weeds. It is reported as particularly bad in the Laikipia district. It has been noted in the north of Tanzania and in the Southern Highlands and occurs in the Kigezi district of Uganda. It is mainly a weed of arable land, especially in small-grain cereal crops, maize and lucerne, and more frequent on the more fertile soils.

Methods of control
Malva is relatively resistant to 2,4-D and MCPA, but appears to be more sensitive to dicamba. Contact chemicals are only effective against young seedlings but various residual chemicals, including members of the substituted urea and triazine groups, can give effective pre-emergence control.

MALVACEAE

Pavonia urens Cav.   Subukia Weed

DESCRIPTION

A large, coarsely hairy, perennial herb or semi-woody shrub growing from 1.5—3 m high with spirally arranged leaves and white (or rarely reddish) flowers about 5 cm across, clustered in the leaf axils. The leaves have a long stalk and a blade about 10 cm long, palmately lobed, indented at the base and with a coarsely toothed margin. The flowers are borne on short stalks, several together in the axils of the upper leaves. Outside the calyx of 5 green-veined sepals is another ring of 10—12 sepal-like bracts; there are 5 overlapping petals, numerous stamens with their filaments joined into a tube and a 5-celled ovary, from which arises a long style, divided at the tip into 10 arms. The calyx persists as the fruit ripens and the ripe carpels into which the fruit divides bear numerous short bristles and 3 long awns.

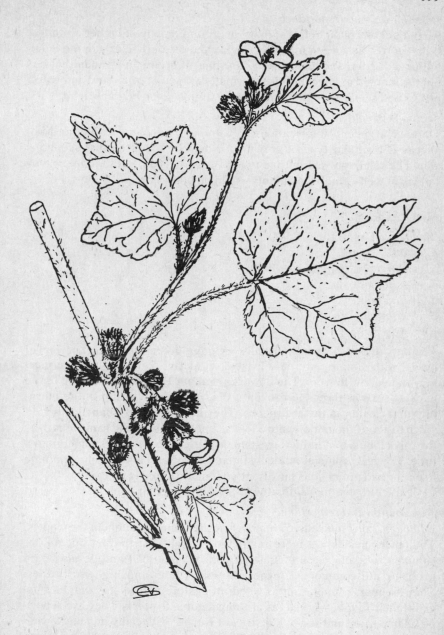

Figure 101. Pavonia ureus

Distribution and importance

An indigenous plant of forest margins, found mostly at higher altitudes. It is reported as a weed in the Kenya highlands, particularly in the Solai-Subukia district, and in northern Tanzania. It occurs in Uganda, but has not been noted as a weed. It is principally a weed of grassland and waste places and seedlings can sometimes be troublesome in arable land.

Methods of control

There is no specific information on *P. urens*. By analogy with other Malvaceae of a similar type of growth, it would be expected that seedlings would be relatively susceptible to MCPA and 2,4-D, but that only a partial kill of well-established plants would be possible.

MALVACEAE

Sida acuta Burm. f.   Prickly Sida
S. alba L.
S. cordifolia L.
S. cuneifolia Roxb.
S. ovata Forsk.
S. rhombifolia L.

DESCRIPTION

A group of low-growing, usually twiggy, green-stemmed, perennial herbs, with simple, alternate leaves, regularly toothed round the margin, and yellow flowers. The flowers are about 1 cm across and are borne in ones, twos or threes in the leaf axils, normally opening in the mornings and closing in the afternoons. They have a saucer-shaped calyx divided to about the middle into 5 lobes, 5 petals, about the same length as the calyx lobes, and numerous stamens with their filaments joined into a tube. The fruit splits at maturity into a ring of carpels, which terminate in one or two projections mostly extended into a fine bristle.

Characters distinguishing the species are given in Table 4.

Distribution and importance

S. alba, S. cordifolia and *S. rhombifolia* are found throughout the tropics. The other species are more restricted in their distribution but all are common in many parts of East Africa. *S. cuneifolia* and *S. acuta* are probably the commonest, especially at medium and higher altitudes. They are mainly important as weeds of pasture and waste land, and because their tough, wiry stems are unpalatable to stock. They are reported as important in the Sotik district of Kenya, especially on heavy 'vlei' soils. *S. ovata* is also principally a pasture weed, while *S. alba, S. cordifolia* and *S. rhombifolia* are more frequently troublesome in arable land.

TABLE 4.
Distinguishing characters of *Sida* species

Species	Maximum height of stems	Leaf size and length of stalk	Leaf shape	Length of flower stalk	Number of carpels	1 or 2 projections on fruit
S. acuta	1 m	5 × 1.2 cm 3 mm	Narrow lanceolate, tip pointed	1.2 cm	5 – 6	2
S. alba	0.6 – m, short spines on stem at base of leaf	4 × 2 cm 12 – 20 mm.	Lanceolate, tip blunt	2 – 2.5 cm jointed	5	2
S. cordifolia	1 m	6 × 4 cm 12 – 25 mm	Heart shaped, surface downy	2 – 4 cm	8 – 10	1
S. ovata	50 cm, densely covered in grey hairs	4 × 2.5 cm 5 – 10 mm	Ovate, tip blunt or rounded	0.5 cm or less	7 – 8	2
S. rhombi-folia	1m, erect little branched	6 × 2.5 cm less than 5 mm	Diamond shaped, under-surface pale	2 – 4 cm. flowers solitary	8 – 12	2
S. cuneifolia	50 – 60cm.	1.2 × 1 cm very short	Wedge shaped, indented at tip	Very short	5	1

Fruit of *S. rhombifolia* is said to be poisonous to poultry.

Methods of control

In grassland, mowing only gives temporary control of *Sida* species and in arable land the deep, woody tap root makes control by cultivation difficult. Young seedlings can be killed by spraying with 2,4-D or MCPA but resistance to these herbicides increases rapidly with age and formulations containing dicamba or picloram are needed to control established plants. Pre-emergence treatment with substituted ureas or triazines is effective in a variety of crops.

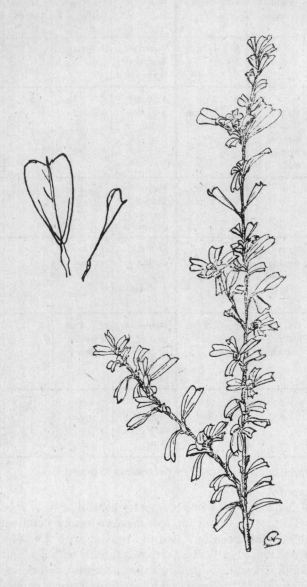

Figure 102. Sida cuneifolia

NYCTAGINACEAE

Boerhavia diffusa L. (= *B. repens* L. var
  diffusa Hook.f. = *B. adscendens* Willd.)   Tar Vine
B. erecta L.
B. coccinea Mill. (= *B. viscosa* Lag. & Rodr.)

DESCRIPTION

A group of perennial herbs with a fleshy rootstock and branches arising
from the base, which are prostrate at first, then ascending. The simple
leaves are in unequal pairs and the small flowers are grouped in umbels
at the ends of the branches of the much branched, loose inflorescence.

In *B. diffusa* the leaves are borne on stalks of very varied length and
are 3 —5 cm long × 2 —4 cm wide, broadly ovate in shape and blunt at
the tip, with a somewhat wavy margin. The under surface of the leaves
is paler than the upper. The inflorescence is leafless and the umbels
terminating the branches are mostly 2 —4 flowered. The flowers are red
and 2 mm across, with a 5-lobed perianth and, usually, 3 stamens. The
fruits are 3 mm long and covered with sticky, glandular hairs, readily becoming detached and sticking to clothing or passing animals.

B. erecta is a more upright plant. The inflorescence is very similar to
the above, but it can be distinguished by the fruits, which lack glandular
hairs and are not sticky.

B. coccinea is also similar in general appearance. The inflorescence is
often leafy. However, the umbels have 4 12 pink or mauve flowers and
the stems are usually hairy and more or less sticky.

Distribution and importance

B. diffusa and *B. erecta* occur throughout the tropics and are widespread in Kenya, Tanzania and Uganda from sea level to about 1500 m.
They are also common in Zanzibar. *B. coccinea* has a more restricted
distribution and is less commonly encountered. The three species occur
mainly as weeds of cultivated and waste land, and are often found in
lawns in drier areas. Although common they are not weeds of outstanding importance.

Methods of control

After mechanical cultivation *Boerhavia* species resprout from the rootstock, but relatively few cultivations are needed to exhaust their powers
of regeneration.

B. diffusa is relatively susceptible to 2,4-D and MCPA and seedlings
are readily controlled. On established plants some regrowth is likely to
occur after a single treatment but retreatment normally results in a

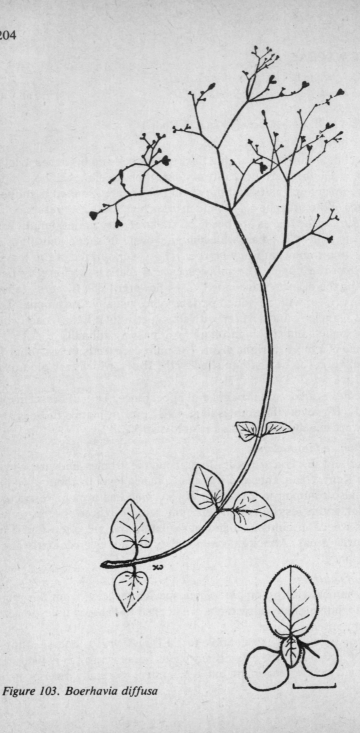

Figure 103. Boerhavia diffusa

complete kill. Atrazine has given good results as a pre-emergence treatment in maize in the Philippines.

OROBANCHACEAE

Orobanche minor Smith   Broomrape
O. ramosa L.   Branched Broomrape

DESCRIPTION

O. minor is a reddish brown, unbranched herb growing to about 60 cm high, with leaves reduced to scales and pale mauve and yellow flowers. It is parasitic on the roots of a wide range of plants, especially members of the Papilionaceae. The stems are often clustered and are covered with soft, whitish hairs, as also are the 2 cm-long scales on the stem. The inflorescence is a spike about half the length of the stem, the sessile flowers arising in the axils of bracts similar to the scales. The flowers are 1.5 –2 cm long, with the petals united into a broadly tubular corolla, which is curved and divided at the tip into a shallowly 2-lobed, upper lip and a 3-lobed lower lip. The calyx is split to the base into 2 sections, each with sharply pointed teeth. There are 4 stamens attached to the corolla tube and an ovary bearing a 2-lobed stigma.

O. ramosa differs in that the stem is often branched and a pale yellow colour, there are one larger and two smaller bracts at the base of each flower instead of a single bract and the calyx is not split into two.

Distribution and importance

O. minor occurs in various parts of Europe, America and North Africa, as well as in East Africa, where it is widespread between 500 and 3000 m. It is often found in gardens and grassland, sometimes in cultivated land, and is reported as common in the Sotik district of Kenya. Serious infestations may develop in red clover and lucerne crops. The minute seed remains viable in the soil for up to 10 years so that, once an infestation develops, it tends to be persistent.

O. ramosa is thought to be an introduction from Europe where it seriously parasitizes crops such as tobacco and tomato, and a wide range of other hosts. It is not common in East Africa but has been recorded in Kenya and northern Tanzania at altitudes around 2000 m and could become more important.

Methods of control

In badly infested areas the reserves of *Orobanche* seed in the soil can be slowly reduced by growing non-susceptible crops for several years and it may be possible to accelerate this process by growing trap crops, such as linseed, which stimulate germination but do not act as hosts. There appear to be some possibilities with herbicides. In Syria, for example, low rates of glyphosate (around 100g/ha) have given a useful measure of control without injuring broad beans and, in Czechoslovakia, maleic hydrazide has been used to suppress *O. ramosa* in tobacco. In the USSR *Phytomyza* flies are giving promising results as a biological control method for certain *Orobanche* species and, in Israel, infestations have been greatly reduced by covering the soil with transparent plastic for 4

Figure 104. Orobanche minor

weeks prior to sowing a crop, thus raising soil temperatures and stimulating germination of the weed in the absence of suitable host plants. With some crops there are also possibilities of developing cultivars resistant to attack by *Orobanche*.

OXALIDACEAE

Oxalis corniculata L.   Yellow Sorrel

DESCRIPTION

A creeping, annual herb with alternate, clover-like leaves and yellow flowers. The stem divides at the base into numerous branches, rooting at the nodes and covered with long hairs. The leaves have slender stalks up to 8 cm long and 3 heart-shaped leaflets, 5 –10 mm long, somewhat broader, and deeply split at the apex. In the evening the leaflets fold downwards around the leaf stalk. The flower stalks, each bearing up to 6 flowers, arise in the axils of the leaves. The stalks of individual flowers are 5 –10 mm long and bend downwards as the fruits develop. The flowers are about 1 cm across with 5 sepals, 5 wedge-shaped petals, 10 stamens and an ovary with 5 stigmas which develops into an elongated capsule. *O. radicosa* A. Rich., a more erect plant with a tuberous rootstock, has been much confused with *O. corniculata* in the past, and is of similar importance as a weed.

Distribution and importance

A cosmopolitan plant of wide occurrence in East Africa, mostly seen at altitudes above 1000 m, but also occurring in Zanzibar. It is most commonly found as a weed in Kenya, and can occur either in grass or arable land. It is troublesome in lawns, where its creeping habit enables it to escape mowing, and has been recorded as a locally important weed of pyrethrum, not only in Kenya, but also in northern Tanzania and the Southern Highlands. In pasture it can cause poisoning under some conditions.

Methods of control

O. corniculata is only temporarily controlled by 2,4-D or MCPA applied as lawn sprays. Formulations containing dicamba are more effective and in the USA, triclopyr has given good control of this species. In cultivated land several residual herbicides can be used to prevent the establishment of seedlings, including metribuzin, diuron, EPTC and oxadiazon.

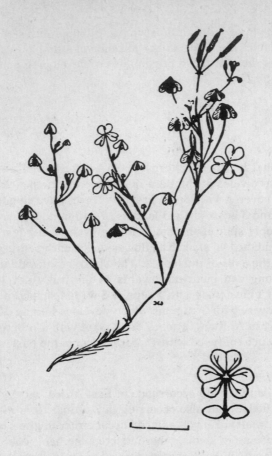

Figure 105. Oxalis corniculata

OXALIDACEAE

Oxalis latifolia H.B.K.   Oxalis
   (including much East African material originally
   identified as *O. semiloba* Sond.)
O. corymbosa DC.

DESCRIPTION

These species are of very different habit to *O. corniculata*, being perennials with a short, thick, tuberous or bulbous rootstock and pink or lilac
flowers. There is no aerial stem, the leaves and flower stalks arising

directly from the rootstock. The leaf stalks are long and slender, with 3 more or less triangular leaflets at the apex.

In *O. latifolia* the leaf stalks are up to 25 cm long and the flower stalks somewhat longer. They arise directly from the scaly bulb and break away from it readily. At the base of the bulb is a white, fleshy, carrot-shaped tap root. The bulb also gives rise to white rhizomes, which grow out to produce small bulbils at their extremities. The leaflets are rather less than 2.5 cm long × more than 2.5 cm wide and often have purplish markings. They have straight edges, rounded corners and an apex indented up to 1/3 the length of the leaflet. The inflorescence consists of 5 —15 flowers arranged in an umbel and individual flower stalks are 1 — 2 cm long. The flowers consist of 5 sepals, each with two orange-coloured glands at its tip, 5 overlapping petals about 1 cm long, 10 stamens joined near the base and an ovary with 5 styles. The fruit is a more or less spherical capsule.

O. corymbosa is similar in general appearance and in its production of numerous small bulbs, but can be distinguished by its more rounded leaflets, often with orange dots, and branched inflorescence.

Several other bulbous *Oxalis* species occur in East Africa and are sometimes confused with the above, especially the true *O. semiloba* (of no significance as a weed) and *O. anthelmintica*. These, however, can readily be distinguished by the possession of a vertical rhizome up to 15 cm long between the base of the leaf stalk and the bulb.

Distribution and importance

O. latifolia is a native of South America, but has been introduced into many tropical and temperate countries and usually appears to become established first in plant nurseries and gardens. It is widely distributed in East Africa, mostly at higher altitudes, being recorded as a serious weed in various parts of the Kenya highlands, especially the Sotik, Elgeyo-Marakwet and Eldoret districts. In Tanzania it is troublesome in the Arusha and Moshi districts and in Uganda is a problem in Toro and the Kawanda district.

It grows best in moist shady situations and, under favourable conditions, such as in the shade of coffee nurseries, the leaves can form a complete ground cover, while the soil becomes a mass of bulbs. It is most commonly found in localized areas in gardens, coffee and tea plantations, etc. but, once established, it appears to be permanent and infested areas gradually extend. In some parts of Kenya and Uganda it also appears to be infesting arable land on a larger scale and, although the leaves die back as soon as conditions become dry, the bulbs are very

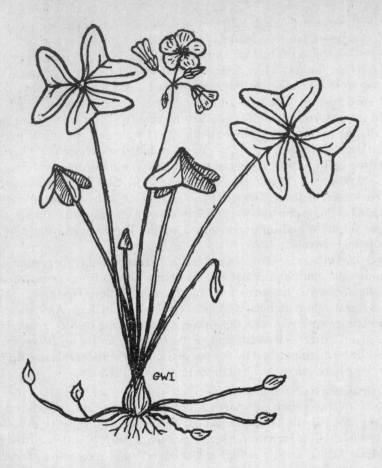

Figure 106. Oxalis latifolia

resistant to desiccation and sprout as soon as moisture becomes available again.

O. corymbosa is also an introduced species. It presents a very similar problem to *O. latifolia,* but so far has only become established at Lyamungu and Amani in Tanzania.

Methods of control

Cultivation has no lasting effect on *O. latifolia* and, even with the most careful removal by hand, numbers of bulbils are invariably left in the soil to grow again as soon as conditions are favourable. One method of control advocated in Britain is to grass the area over for at least 3 years, but a vigorous sward is necessary to prevent the growth of the weed and it is not certain that, even after 3 years, some bulbils will not be able to grow again when the grass has been ploughed up.

The foliage is readily killed by many herbicides, but few foliar-acting chemicals are translocated sufficiently to reach the bulb system, and merely killing the leaves provides a stimulus to bulbil ripening and a consequent increase in the number of bulbils that are viable. Aminotriazole is translocated more readily and will kill a proportion of the bulbils, so that repeated treatment with this chemical, with occasional cultivation to stimulate more dormant bulbils into growth, may eventually result in eradication. Repeated treatment with glyphosate is also a possibility.

Of the residual herbicides only oxadiazon shows much potential against *Oxalis*. Soil sterilization is an expensive alternative. In New Zealand methyl bromide has been used successfully to eradicate a very similar species of *Oxalis* and in Britain metham-sodium has also been employed successfully against *O. corymbosa*.

PAPAVERACEAE
Argemone mexicana L.

Mexican Poppy
Mexican Thistle

DESCRIPTION
A much branched, prickly, annual herb growing to 1 m high, with alternate leaves and showy yellow flowers. The plant exudes a yellow juice when cut. The blue-green leaves are sessile and more or less sheathing at the base, up to 15 cm or more long and pinnately lobed, with an irregularly serrated, sharply spiny margin. The veins of the leaves are outlined in greyish white on the upper surface. The flowers are up to 5 cm across and borne at the ends of the branches. They consist of 3 prickly sepals, 6 petals, numerous stamens and a capsule of 4 —6 divisions, topped by a short persistent stigma. The fruit is spiny, up to 4 cm long × 2 cm across and opens at the tip by a number of slits releasing numerous, brown-black seeds 2 mm in diameter.

Distribution and importance
A native of Central America which has been introduced into most tropical and sub-tropical countries. It was recorded in Zanzibar in 1860 and is now distributed widely through East Africa, especially in Tanzania, at altitudes up to 1800 m, both in waste ground and arable land. It is a particularly serious problem in northern Tanzania in maize, beans and other arable crops. It appears able to germinate throughout the year, even during the dry season, and often develops during the later stages of crop

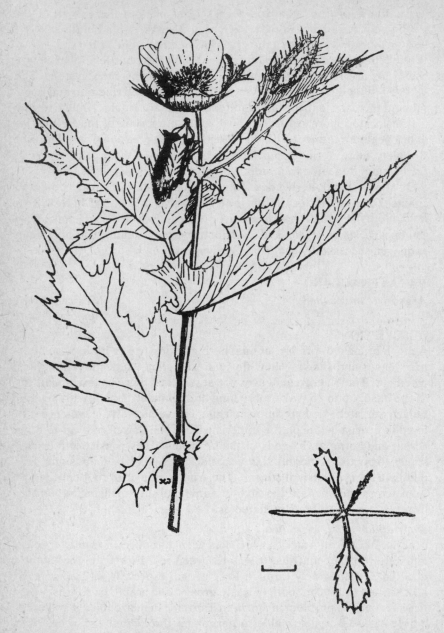

Figure 107. Argemone mexicana

growth so that it constitutes a hindrance to harvesting. The seed is poisonous to poultry.

Methods of control
Argemone is only moderately susceptible to 2,4-D and MCPA, but small seedlings less than 5 —8 cm high can be satisfactorily controlled with rates of the order of 2 kg per ha and pre-emergence treatment can also be effective. Improved control of established plants is given by formulations containing dicamba. Contact herbicides are relatively ineffective, as is pre-emergence treatment with atrazine or diuron. At Arusha, however, promising results have been obtained with metribuzin and mixtures of alachlor and linuron.

Although some measure of chemical control is possible in the early stages of growth of various crops, the problem of seedlings becoming established in the later stages has not yet been satisfactorily solved.

PAPILIONACEAE

Crotalaria chrysochlora Harms   Rattlepod
C. incana L. subsp. *purpurascens* (Lam.) Milne-Redh.
C. brevidens Benth. var. *intermedia* (Kotschy) Polhill
C. laburnifolia L.
C. polysperma Kotschy

DESCRIPTION
There are nearly 200 species of *Crotalaria* in East Africa, many of which can be regarded as weeds in some situations. It is impossible to give descriptions of all those likely to be encountered, but the five species described are among the commonest.

They are all annual or short-lived perennial herbs with alternate leaves consisting of 3 leaflets with entire margins (some other *Crotalaria* species have simple leaves) and shortly stalked flowers borne in an erect, elongated, unbranched terminal inflorescence. The flowers are usually yellow (sometimes blue) and of typical pea-flower formation, with 10 stamens united into a tube, which is split in the upper part. The fruit is a more or less inflated pod, in which the seeds rattle loosely when ripe.

Characters distinguishing the five above mentioned species are given in Table 5.

Distribution and importance
C. incana is of widespread tropical distribution, *C. brevidens* occurs also in tropical America, the other species are African. They are widely distributed in East Africa in grassland of various types, *C. incana* sometimes occurring also in cultivation and *C. laburnifolia* being found main-

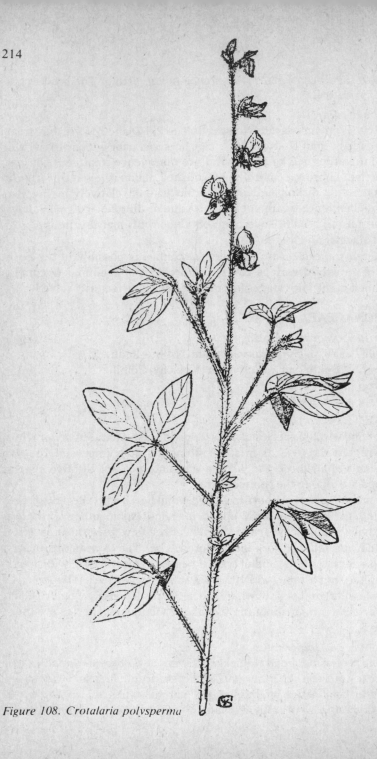

Figure 108. Crotalaria polysperma

TABLE 5.
Distinguishing characters of *Crotalaria* species

Species (A = annual P = perennial)	Flower colour and length of keel	Dimensions of largest leaflet and length of stalk	Length of pod	Other characters
C. chrysochlora (P) Semi-prostrate, 10-15 cm. high.	Yellow with red or purple veins 5-10 mm	3 × 1 cm (hairy beneath) 5-10 mm	1 cm (hairy, stalk short)	Inflorescence rather dense
C. incana (A) 0.6-1.2 m. high	Yellow, with purple veins, 1 cm	5 × 2 cm 1-2 cm	2.5-4.5 cm (with coarse grey hairs)	Inflorescence loose
C. brevidens (P) 0.6-1.2 cm. high	Pale yellow or whitish with purple veins, 2-2.5 cm	15 × 2 cm (short hairs beneath) 2-4 cm	5 cm (soon becoming hairless)	Inflorescence long, more or less interrupted
C. laburnifolia (P) 1-1.5 cm. high	Yellow 2.5-3 cm	5 × 2.5 cm (hairless) 4-8 cm	5 cm (hairless, stalk long)	Inflorescence loose, whole plant hairless
C. polysperma (P) 0.6-1.2 m. high	Violet-blue 1.5-2 cm	5 × 2 cm (silky on both sides) 2-3 cm	4-5 cm (with rusty hairs, stalk short)	Inflorescence with about 6 flowers, plant covered rusty hair.

ly in drier areas. A number of *Crotalaria* species are poisonous to stock. *C. incana* subsp. *purpurascens* and *C. polysperma*, for example, are known to be poisonous, while various other species are suspect. *C. brevidens* var. *intermedia*, however, is non-toxic and a useful constituent of grazing land. Certain species of the genus contain substances which have proved to cause liver cancer in humans.

Methods of control
The annual *C. incana* is susceptible to 2,4-D and in Trinidad has been controlled in groundnuts with MCPB. Little information is available on the susceptibility of the perennial species in East Africa, but various perennial *Crotalaria* in other parts of the world are reported to be controlled effectively with 2,4-D or combinations of 2,4-D with dicamba etc.

216

PAPILIONACEAE

Indigofera spicata Forsk.   Creeping Indigo
   (= *I. hendecaphylla* Jacq.)

DESCRIPTION

A tough-stemmed, more or less creeping, perennial herb with a deep tap root, alternate, pinnate leaves and spikes of small, red flowers. The greyish leaves are up to 8 cm long, with one terminal and 4 —8 alternately arranged, lateral leaflets. These are up to 2.5 cm long, sessile and lanceolate, with a pointed tip and smooth margin. The flowers are borne in dense spikes which arise from the leaf axils and are longer than the leaves. Individual flower stalks are very short, the flowers about 8 mm long, with a 5-pointed calyx, 5 unequal-sized petals, the lower two joined to form a keel, and 10 stamens, 9 joined together and one separate. The fruits are narrow, cylindrical pods, up to 2.5 cm long, directed downwards and containing 4 —8 seeds.

Distribution and importance

A plant found in many parts of the tropics, which is widely distributed throughout Kenya and Tanzania (including Zanzibar) over a wide range of altitudes. It is reported as a weed of some importance in the Sotik and Molo districts of Kenya and in the Southern Highlands of Tanzania. It occurs both in grassland and arable crops and is a common constituent of lawns, where it is unaffected by mowing because of its prostrate growth. It also occurs commonly as a pioneer plant on soil bared by erosion. It was at one time considered as a potentially useful legume for pasture improvement, but has been found to be toxic to stock and chickens, the toxic principle being ß -nitropropionic acid.

Methods of control

I. spicata is of limited susceptibility to 2,4-D and MCPA, but it can be killed in the young seedling stage. With established plants the tops are killed back, but there is often a considerable amount of regrowth and repeated treatment is needed to achieve a complete kill. As with several other lawn weeds belonging to the family Papilionaceae it is probable that formulations containing dicamba may prove more effective.

PAPILIONACEAE

Mucuna pruriens (L.) DC.   Buffalo Bean

DESCRIPTION

A tall growing, weak-stemmed, hairy, annual climber, with alternate, trifoliate leaves, dark purple flowers and pods densely covered with

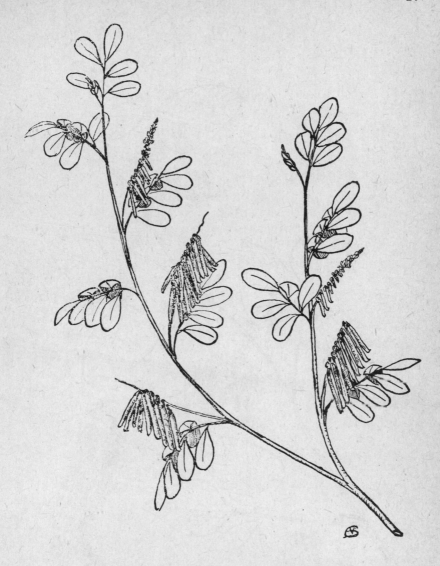

Figure 109. Indigofera spicata

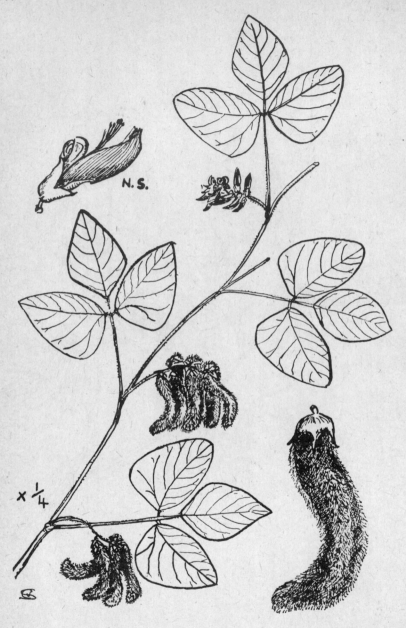

N.S.

× 1/4

Figure 110. Mucuna pruriens

orange hairs which cause intense irritation if they touch the skin. The leaves have long stalks and the 3 leaflets are up to 15 cm long × 5 cm wide, sharply pointed at the tip, with the lateral leaflets very asymmetrical and the under surfaces covered with grey, silky hairs. There are two small leafy projections at the base of the stalk of each leaflet. The inflorescences arising in the leaf axils are shortly stalked and few-flowered, the stalk being swollen at the nodes where the flowers arise. The flowers are about 4 cm long, the pods curved and up to 9 cm long × 1 cm broad, with longitudinal ribs. When the pods are ripe, the irritant hairs are dislodged by very slight movements, so that passing near the plant is often sufficient for a painful dose of hairs to be received.

Distribution and importance
An indigenous plant found mostly at lower altitudes in the coastal provinces of Kenya, in the Tanga district and in the east and south of Tanzania and in Zanzibar. It does not appear to be common in Uganda. It is principally a plant of bush land, overgrown plantations, etc., but also occurs quite commonly in sisal in Tanzania, where it is particularly objectionable and can seriously hinder harvesting.

Methods of control
In crops where *Mucuna* is known to exist, control measures should be applied in the early stages before the pods develop. Isolated plants can readily be hoed out, or larger patches can be sprayed with 2,4-D or MCPA, to both of which herbicides it is susceptible. Once the pods have matured no control measures are possible without the risk of considerable discomfort.

PHYTOLACCACEAE

Phytolacca dodecandra L'Herit.

DESCRIPTION
A perennial climber or sprawling, thicket-forming shrub, which is woody at the base, arises from a large fleshy tuber and has green, somewhat succulent stems, alternate, simple leaves and long, cylindrical spikes of small white or greenish flowers. The older branches are marked with prominent leaf scars. The leaves are up to 15 cm long × 10 cm wide, pointed at the tip and rounded at the base, with smooth, translucent margins and stalks up to 5 cm long. The inflorescences arise in the leaf axils and are 30 cm or more long × 2 cm diameter, with numerous, shortly stalked flowers. The flowers have 4—5 sepals, but no petals. Male flowers have about 15 stamens, female flowers usually 5 carpels developing into small, orange or red fruit.

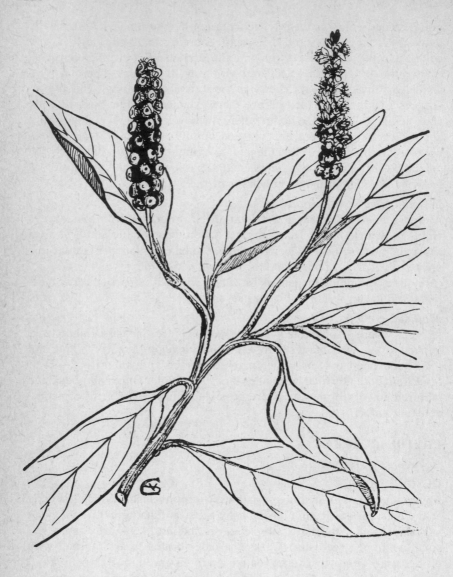

Figure 111. Phytolacca dodecandra

Distribution and importance
Phytolacca is widely distributed in Kenya, Uganda and Tanzania, mostly in the areas of higher rainfall. It occurs mainly on the edges of forest and thickets and is common as a weed of waste ground and hedges. It is important because it is poisonous to cattle and other animals. A potentially useful molluscicide is found in the fruit.

Methods of control
Repeated cutting gives control, but it is often difficult to locate the base of the stem. Little is known about its reaction to herbicides. A related species in New Zealand is listed as susceptible to 2,4-D when young and, applied under favourable growing conditions, dicamba or picloram formulations are effective on older plants. Glyphosate is another chemical which would be expected to give control.

POLYGONACEAE

Emex australis Steinh.   Devil's Thorn
E. spinosus (L.) Campd.

DESCRIPTION

Distinctive, weak-stemmed, annual herbs growing 0.6 —1 m high, with alternate or clustered, simple leaves, separate male and female flowers, both types greenish and inconspicuous, and fruits with 3 very sharp spines.

E. australis has leaves with narrow stalks 8 cm or more long and blades up to 8 cm long × 5 cm broad, abruptly narrowed to the blunt or rounded tip and abruptly narrowed or with rounded lobes at the base. The flowers are borne in terminal and axillary clusters, the male flowers very small with 5 —6 perianth segments and 4 —6 stamens, the female flowers lower down the stem with a perianth of 6 segments, which grows larger as the fruit ripens within and eventually reaches 5 mm long × 10 mm across. The 3 inner segments are upright and terminate in fine bristles, the 3 outer ones are joined into a tube below and extended at the tips into spreading, rigid spines 5 mm long.

E. spinosus is very similar in general appearance, but can be distinguished by the smaller, differently shaped fruit, 5 mm long × 4 mm across, with smaller 2 —3 mm spines and a projecting rim at the base of the perianth tube.

Distribution and importance
Both species have been introduced into East Africa, *E. australis* from South Africa, *E. spinosus* possibly from the Mediterranean region. So far they are restricted to the Rift Valley and Central Provinces of Kenya

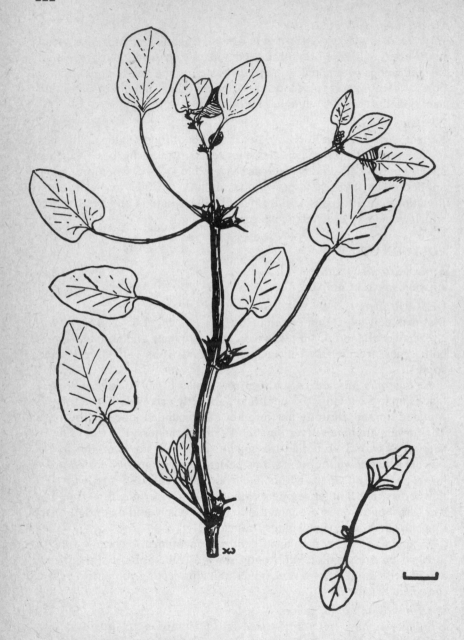

Figure 112. Emex australis

at altitudes between 1500 —1800 m, especially in the Uasin Gishu and Kiambu districts. *E. australis* is the commoner of the two. They occur mostly as weeds of waste land, along railways, etc., but can also occur in arable land, and in Australia *E. australis* is a serious weed of cereal crops. If the plants are grazed by stock the spiny fruits can cause much trouble and oxalic acid in the leaves can cause poisoning.

Methods of control
The *Emex* species are of limited susceptibility to 2,4-D and MCPA. Dicamba is more effective and kills plants at a later stage of growth. Prometryne is a recommended treatment against young seedlings in legume pasture in Australia and linuron is also effective.

POLYGONACEAE

Oxygonum sinuatum (Meisn.) Dammer   Double Thorn
  (= *O. atriplicifolium* var. *sinuatum* (Meisn.) Bak.)

DESCRIPTION
A much branched, sprawling, annual herb, growing 30 —60 cm high with alternate leaves, inconspicuous white or pink flowers borne in long, slender, unbranched, spike-like inflorescences and fruits with 3 short prickles. The leaves are up to 5 cm long × 2.5 cm wide on 1 —2 cm stalks, more or less deeply divided into a few irregular lobes, especially in the lower half, and narrowed gradually to each end. Where the leaf stalk joins the stem it is expanded into a reddish, tubular portion clasping the stem for a length of about 5 mm. The inflorescences are up to 30 cm long and arise in the axils of one or more of the upper leaves while the flowers arise in groups of 2 —4 in the axils of tubular bracts, the lower ones 2 —5 cm apart. Flowers are about 3 mm across, with a tubular, 4 —5 lobed perianth, 8 stamens and an ovary with 3 styles. The fruits are about 5 mm long pointed at each end and extended into 3 short, spreading prickles in the middle.

Distribution and importance
An indigenous species confined to Africa and occurring widely in all parts of East Africa from sea level to 2000 m. It is recorded as a serious weed in the Machakos and Uasin Gishu districts of Kenya, but is also common as an arable weed in most other cultivated areas and in most types of crop. Apart from competing with crops for moisture, light and nutrients, its thorny seeds can cause injury to the feet of man and animals.

Methods of control
Young seedlings can be effectively controlled with 2,4-D and good re-

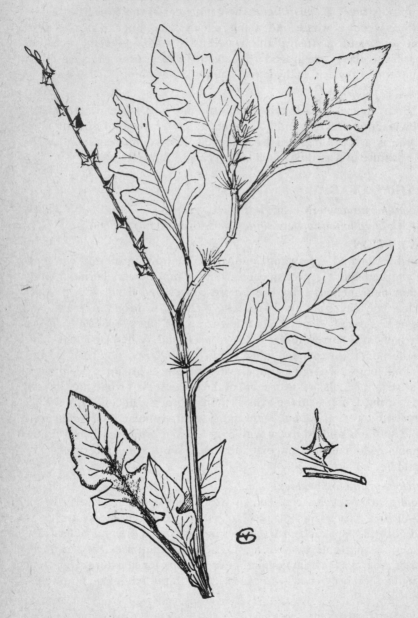

Figure 113. Oxygonum sinuatum

sults have also been obtained with 2,4-DB and dicamba formulations. On its own, however, MCPA appears to be less effective and MCPB has failed to give control. Treatments should be applied as early as possible, as the plants rapidly increase in resistance with age. Effective control by pre- or early post-emergence treatment with residuals, including atrazine, diuron, pendimethalin and alachlor is reported and the weed is very sensitive to the contact action of paraquat.

POLYGONACEAE

Polygonum aviculare L.	Knotgrass
P. persicaria L.	Redshank
P. nepalense Meisn.	
(= *P. alatum* Spreng.)	
P. amphibium L.	Perennial Knotgrass

DESCRIPTION

A group of weak-stemmed, herbaceous weeds, with alternate, simple leaves, expanded at the base of the stalk into a short, tubular sheath around the stem, and small pink or whitish flowers in axillary clusters or dense, cylindrical spikes.

P. aviculare is a prostrate annual with a tough tap root and wiry branches. The leaves are very variable in size, up to 5 cm long × 1.2 cm wide, gradually narrowed to the rather blunt tip and to the short stalk and the sheaths are silvery. The numerous, inconspicuous flower heads arise in the leaf axils and consist of up to 6 crowded flowers, 3 mm long on short stalks. There is a perianth of 4—5 segments joined at the base, 5—8 stamens and 3 styles. The one-seeded, reddish brown fruits are 3 angled and 3 mm long.

P. persicaria is an annual with somewhat fleshy, green or red stems rooting at the lower nodes, but later becoming erect. The leaves are short stalked, up to 10 cm long × 2 cm wide, narrowed gradually at each end and often with a black blotch on the upper surface. The stout, 2 cm long flower heads are both terminal and axillary on stalks of variable length, and consist of numerous pink flowers.

P. nepalense is also a straggling annual, with relatively broad leaves up to 5 cm long × 3 cm wide which are abruptly narrowed to the pointed tip and to a broadly winged stalk up to 1 cm long. The flower heads are mainly terminal, shortly stalked and 5 mm across, each containing about 12 flowers.

P. amphibium is similar to *P. persicaria* but larger and a perennial with stout stems and an extensive rhizome system. The leaves are up to

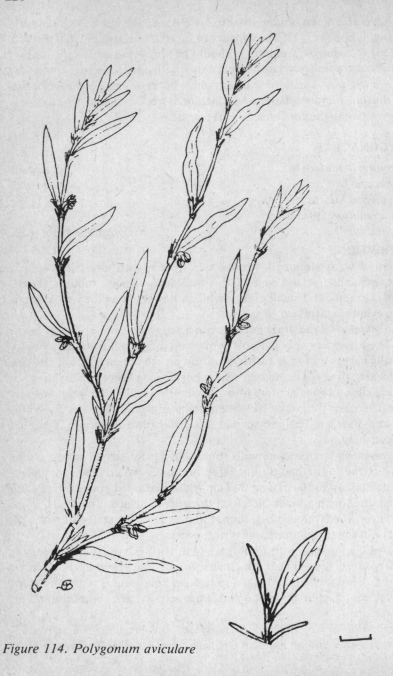

Figure 114. Polygonum aviculare.

15 cm long × 3 cm wide, pointed at the tip and rounded at the base, with stalks up to 1 cm long. The flower heads are mostly terminal up to 4.5 cm long × 1 cm in diameter, and are borne on slender stalks.

Distribution and importance
With the exception of *P. nepalense* all the species have been introduced, probably from Europe, and are of limited distribution in East Africa. *P. aviculare* is common in arable land in parts of the Kenya highlands at about 2500 m, especially on the Kinangop, and appears to be spreading. *P. persicaria* occurs in the Busoga district of Uganda as a weed of cultivated ground, while *P. amphibium*, again, occurs in the Kenya highlands, either in cultivation or in wet places. The two latter species are, at present, only of very limited importance as weeds, but are well established and probably spreading.

   P. nepalense is more characteristic of the tropics and more widely distributed, having been recorded in all three East African countries at altitudes of 1400 −2400 m. It sometimes occurs as a weed in cereal crops, especially on the Kinangop and in the Mweiga and Naro Moru areas of Kenya.

Methods of control
East African experience agrees with findings in Europe that MCPA is of little use against *P. aviculare*, while 2,4-D is only effective on young seedlings. Dichlorprop may be somewhat better but herbicide formulations containing bromoxynil, ioxynil or dicamba are needed to obtain reliable control of established plants. Among residual chemicals *P. aviculare* is only moderately susceptible to diuron and atrazine but ametryne, metribuzin and trifluralin appear to be effective.

   P. persicaria reacts to herbicides in a similar manner to *P. aviculare* in Europe but is more susceptible to the effects of dichlorprop. *P. amphibium* is more difficult to control and is only checked by 2,4-D. *P. nepalense* is reported to be susceptible to 2,4-D in India, and in southern Tanzania has been found more susceptible to diquat than paraquat.

POLYGONACEAE

Polygonum convolvulus L.   Black Bindweed

DESCRIPTION
P. convolvulus differs from the *Polygonum* species described in the previous section both in its habit of growth and in its more widespread occurrence as a weed. It is an annual, climbing plant with weak, twining stems. The leaves are alternate, up to 5 cm long × 4 cm wide, with a stalk up to 4 cm long and a heart-shaped blade, which is pointed at the

Figure 115. Polygonum convolvulus

tip, indented at the base and mealy on the lower surface. The sheath where the leaf stalk joins the stem is brown and split deeply into narrow segments. The inflorescences arising from the leaf axils may be sessile or stalked and consist of clusters of very shortly stalked greenish flowers which are about 3 mm long. The perianth is divided into 5 segments, the outer 3 becoming narrowly winged as the fruit develops; there are 8 stamens and the black fruit is 3-angled.

Distribution and importance

Another species which has been introduced into Kenya from Europe, *P. convolvulus* is a common arable weed in many parts of the highlands between.1600 —2000 m and is reported as one of the more important annual weeds in the Eldoret, Laikipia and Naivasha districts. It is particularly important in cereal crops, where it can have a smothering effect and where its twining growth encourages lodging and hinders harvesting. The seed is recorded as causing injury if it is eaten by fowls.

Methods of control

Like the previous species *P. convolvulus* is only moderately susceptible to 2,4-D and 2,4-DB. It is effectively controlled by dichlorprop, however, and also by mixtures containing bromoxynil, ioxynil or dicamba. Contact herbicides give a good kill of young seedlings and control of germinating seeds is also possible with such soil-applied herbicides as simazine, atrazine, diuron and trifluralin.

POLYGONACEAE

Rumex acetosella L. (= *R. angiocarpus* Murb.)	Sheep's Sorrel
R. crispus L.	Curled Dock
R. abyssinicus Jacq.	Dock
R. bequaertii De Wild.	

DESCRIPTION

A group of perennial herbs; *R. acetosella* slender and up to 30 cm high, the other species thick-stemmed and reaching 1 m or more. The leaves are alternate, the flowers small, greenish or reddish and very numerous in whorls along the branches of a terminal inflorescence.

R. acetosella has an extensive underground rhizome system and the leaves are crowded towards the base of the stems. The leaf stalks are up to 6 cm long, the blades up to 5 cm long × 2 cm wide, with a rather bluntly pointed tip and two narrow, basal lobes, curving outwards and giving the leaf a more or less arrow-shaped outline. The inflorescence is leafless, about 15 cm long, and made up of whorls of very small flowers. The flowers (whose structure is more easily seen in the larger species)

Figure 116. Rumex acetosella

have a perianth of 6 segments in two rings of 3, the outer ones small and thin, the inner ones surrounding the ovary, much larger and expanding with the 3-angled fruit.

R. crispus has a deep tap root and grows to 1 m high. The leaves are up tó 25 cm long × 6 cm wide, narrowed to a blunt tip and narrowed or rounded at the base, with a wavy and much curled margin. The stalks of the basal leaves are up to 20 cm long, those of the stem leaves only about 2 cm. The inflorescence is little branched and dense, with the whorls of flowers close together. The fruits are brown and 3 mm long.

R. bequaertii is similar in general appearance to *R. crispus*, but taller, reaching 1.5 m or more, with longer, flatter leaves and an inflorescence with longer branches from which the flowers hang downwards. The clearest distinguishing character is that the inner perianth segments are brown and have 5 or 6 long, hooked teeth on each side.

R. abyssinicus is larger again, up to 3.5 m high, and has long-stalked leaves up to 30 cm long × 20 cm wide, varying from broadly arrow-shaped to strap-shaped with diverging, long, narrow, basal lobes. The inflorescence is large and much branched and the flowers are borne on slender stalks up to 5 mm long.

Distribution and importance

R. acetosella and *R. crispus* are natives of temperate regions and have been introduced into Kenya, probably from Europe. *R. acetosella* is commonest in the Nakuru district at 2000 – 2500 m and is a weed both of arable land, especially cereal crops, and of grassland, being favoured by poor, acid soils. *R. crispus* occurs mainly in the Nairobi area as a weed of damp pastures and waste land. The other two speceies are indigenous and are more widely distributed in most parts of East Africa at altitudes from 1000 –3000 m. They are principally weeds of upland grassland and are commonly reported as troublesome in pastures.

Methods of control

In the seedling stage *R. acetosella* can be controlled with 2,4-D but established plants are resistant. Propyzamide can be used effectively against this species in some situations and amitrole gives control on uncropped ground. Established *R. crispus* also tends to recover from 2,4-D treatment but dicamba formulations are more effective and asulam gives good results on this species. On tall-growing docks in grassland glyphosate applied by wiper-bar is an effective treatment. There is little information on the indigenous species though a mixture of MCPA and 2,3,6-TBA is reported to give somewhat better control of unspecified East African docks than 2,4-D or MCPA alone, and dicamba would be expected to be equally effective.

POLYPODIACEAE

Pteridium aquilinum (L.) Kuhn   Bracken Fern

DESCRIPTION

A fern with an extensive system of underground rhizomes from which the erect, solitary, much divided fronds grow to heights of 2 m or more. As the fronds emerge above ground they are covered with brown scales and at first are coiled into a spiral which straightens as the frond grows. The frond stalk is about the same length as the blade, which is triangular in outline, up to 60 cm wide and pinnately divided. The branches themselves are divided pinnately and the branchlets are pinnately lobed or divided into broad-based segments, up to about 1 cm long × 5 mm wide, bluntly pointed at the tip and with the margins curled downwards. The spores are borne round the edges of the segments on the underside.

Distribution and importance

A species of world-wide distribution, extending from the tropics to the Arctic and occurring commonly in East Africa in areas of higher rainfall. It is found in all three countries, mainly in or near forest, at altitudes above 1500 m, but also occurring commonly in Zanzibar and Pemba. It attains some importance in Kenya as a weed of pastures at higher altitudes. It can form very dense infestations and greatly reduce the grazing potential. Although normally avoided by cattle it is grazed under some circumstances and can cause poisoning.

Methods of control

There is little information about the control of *Pteridium* in East Africa. In some other parts of the world intensive pasture development, including such measures as crushing, burning, application of fertilizer, oversowing and trampling by livestock, have given effective control. It is relatively easy to kill back the fronds with growth-regulator herbicides but difficult to prevent regrowth from the rhizomes. Picloram is the most effective of the chemicals of this type. A chemical more widely used for bracken control, especially in the context of plantation forestry, is asulam and glyphosate can also be used successfully in some situations.

PORTULACACEAE

Portulaca oleracea L.   Purslane
P. quadrifida L.

DESCRIPTION

P. oleracea is a sprawling, fleshy-stemmed, annual herb with numerous, often reddish, branches up to 30 cm long, spirally arranged or opposite, shiny leaves, which are simple and often crowded towards the ends of

Figure 117. Pteridium aquilinum

the branches, and yellow flowers about 1 cm across, only opening fully in bright sun. The thick leaves are up to 3 cm long × 1 cm across, broadest towards the tip, gradually narrowed to a stalkless base and rounded at the tip. The sessile flowers are terminal, or sited at the forkings of the stems, and are solitary or in groups of up to 5, surrounded by a clustered ring of leaves. There are 2 fleshy, unequal sepals, joined towards the base, 4 —5 petals, 7 —12 stamens and an ovary with a short style divided into 3 —6 branches. The fruit is a capsule about 5 mm long containing numerous small seeds.

P. quadrifida is very similar in general appearance, but differs in the paired, ovate leaves, pointed at the tip, in the presence of a ring of long hairs round the stem between each pair of leaves and in the solitary flowers with 4 petals and a 4-branched style.

Distribution and importance

A very common arable weed throughout tropical, sub-tropical and warm-temperate areas, *P. oleracea* is widespread and common in all parts of Kenya, Tanzania and Uganda. In Kenya it is recorded as a particularly bad weed in the Machakos district. It occurs in all types of crop and is especially numerous in irrigated crops and gardens. Large quantities of seed are produced and dense infestations often arise. *P. quadrifida* is also widely distributed, but is less common as a weed and rarely occurs in such large numbers.

Methods of control

The fleshy stem of *P. oleracea* enables it to remain living for several days after the root has been cut and it very often becomes re-established after hoeing. In general, it is not very susceptible to growth-regulator type herbicides. 2,4-D and MCPA can be effective as pre-emergence sprays and on very young seedlings, but are ineffective against established plants. Formulations containing dicamba appear to give better control. With such contact herbicides as bentazon and paraquat also good results are only obtained on young seedlings, older plants rapidly becoming resistant.

In contrast to its relative resistance to foliar-applied herbicides, *P. oleracea* is sensitive to a wide range of residual chemicals. Good control is reported with pre- or early post-emergence application of chemicals belonging to the subsituted urea, triazine, carbamate, amide and dinitro-aniline groups.

Much less information is available about *P. quadrifida*. In the Sudan it is reported to be relatively resistant to 2,4-D and MCPA.

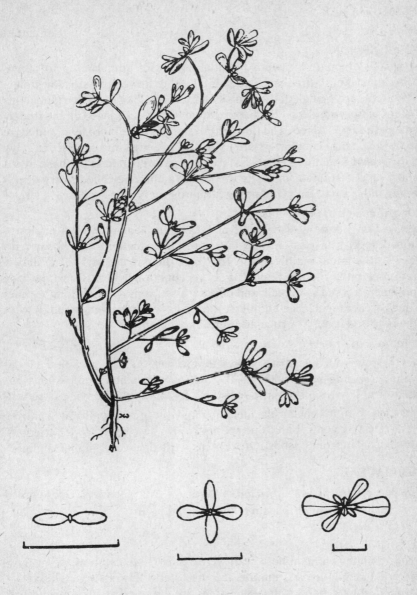

Figure 118. Portulaca oleracea

PRIMULACEAE
Anagallis arvensis L.   Pimpernel
DESCRIPTION
A weak-stemmed, sprawling, annual herb with branches up to 60 cm long, opposite, entire leaves and single, blue flowers on slender stalks. The stems are 4-angled, the leaves up to 2.5 cm long × 1 cm wide, pointed at the tip, rounded at the base, dotted with black glands on the undersurface and unstalked. The flower stalks are 2 cm or more long and arise in the leaf axils. The flowers are up to 1.2 cm across and consist of a calyx with 5 narrow, pointed lobes, 5 spreading petals joined at the base into a short tube, 5 stamens opposite the petals and a one-celled ovary developing into a small fruit containing numerous very small seeds.

Distribution and importance
Anagallis is widely distributed in temperate regions and is an introduced species in East Africa. Like most introductions from cooler regions, it is found mainly in the Kenya highlands above 1800 m and is common in the Eldoret region. It also occurs in northern Tanzania. It is primarily a weed of cereals and other arable crops, but, although sometimes occurring in large numbers, it does not compete very strongly with the crop and is not a particularly serious problem.

Methods of control

Experience in East Africa shows that *Anagallis* can be killed in the young seedling stage with 2,4-D, MCPA or dicamba formulations but that it rapidly becomes more resistant, so that early spraying is recommended. Contact chemicals, including ioxynil, bromoxynil and bentazon give effective control and this species is also susceptible to a variety of residuals, including substituted ureas, triazines and dinitroanilines.

RUBIACEAE

Galium spurium L. var. *africanum* Verdc.   Cleavers, Goose grass
(= var. *echinospermum* auct. non
(Wallr.) Desportes)
DESCRIPTION
A scrambling, annual herb with weak, prickly stems up to 1.2 m long, whorled simple leaves, minute greenish white flowers in small axillary groups and 2-seeded fruits. The stems are 4-angled and the downwards pointed prickles are borne on the angles. The leaf whorls consist of 6 —8 narrow leaves up to 5 cm long and abruptly narrowed at the tip into a short point. The flowers are 5 mm across and are borne 3 —9 together on

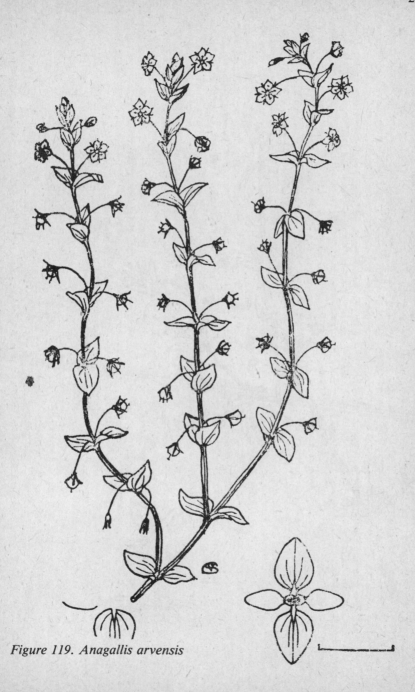

Figure 119. Anagallis arvensis

Figure 120. Galium spurium

stalks of varying length in branched clusters arising in the axils of paired, leaf-like bracts. The flowers have 4 petals, joined at the base into a
very short tube, 4 stamens and an ovary with 2 styles, developing into a
blackish, 2-celled fruit, 3 mm across and covered with hooked hairs.

Distribution and importance

A weed principally of European and North African distribution and probably introduced into East Africa, where it occurs at higher altitudes. It is
found in the Kigezi district of Uganda, in northern Tanzania and the
Usambara and Uluguru mountains, but is most widespread in the highlands of Kenya, being recorded as very serious in Laikipia and one of the
more important weeds in the Naivasha district. It can occur in a wide
range of arable crops, but is most troublesome in cereals, where it encourages lodging, impedes harvesting and sometimes completely
smothers the crop.

Methods of control

Like its close relative *G. aparine* in Britain, *G. spurium* is unaffected by
2,4-D or MCPA. Mecoprop, however, gives good control and formulations of mecoprop with dicamba are very effective. *G. aparine* tends to
recover from the effects of contact herbicides but paraquat has given
good results on *G. Spurium* in tea in southern Tanzania. *Galium* species
are little affected by many of the residual chemicals but pendimethalin is
effective as a pre-emergence treatment.
treatment.

RUBIACEAE

Richardia brasiliensis Gomes
R. scabra L.

DESCRIPTION

R. brasiliensis is a tap-rooted annual or short-lived perennial herb with
spreading, prostrate branches up to 40 cm long, bearing opposite pairs
of leaves and terminal heads of small white flowers subtended by 2 unequal-sized pairs of sessile leaves. The stems are branched and covered
with long, white hairs. The leaves are entire, up to 6 cm long × 2.5 cm
wide, pointed at the tip and narrowed gradually at the base into a short
stalk. Pairs of leaves are joined at the base by stipules up to 3.5 mm long
and fringed by several long narrow projections. There are short hairs all
over the upper leaf surface. The flower heads are up to 1.2 cm in diameter and the 2 pairs of associated leaves have several parallel veins
arising from the base. Individual flowers are sessile, with a 5—6 lobed
calyx, the lobes becoming reflexed in fruit, and a tubular corolla about

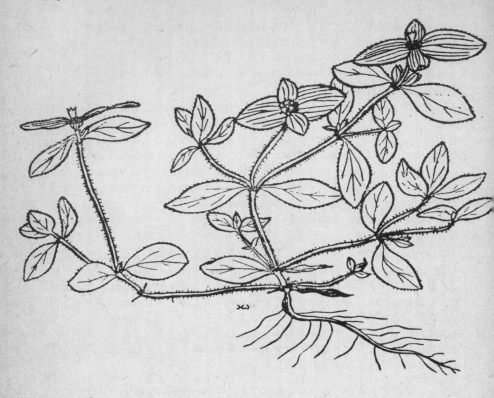

Figure 121. Richardia brasiliensis

3 mm long borne on top of the 3 or 4-sectioned ovary.

R. scabra is a very similar plant but can be distinguished by the fact that the upper surface of the leaf is only hairy near the edges and by the calyx lobes remaining erect in fruit. Another difference is that in *R. scabra* the area joining the ovary sections is narrow, leaving a narrow scar when the fruit splits apart, whereas in *R. brasiliensis* the junction is wide and the scar covers most of one surface of the sections of the fruit.

Distribution and importance

The *Richardia* species originated in tropical America but have been introduced into other tropical and sub-tropical regions and are now distributed widely. *R. brasiliensis* is common in many parts of Africa and is the more frequent species in East Africa, having been recorded from the coast up to 2000 m in Kenya, Tanzania, Uganda and Zanzibar. *R. scabra* is a common species in the southern USA but in East Africa has only been found in Tanzania, where it occurs over the same range of altitudes. Both species are weeds of cultivated land and can be important in maize and various other annual crops. *R. brasiliensis* is also a common weed of coffee in Kenya.

Methods of control
A variety of chemicals is available for the control of these two species. *R. brasiliensis,* for example, is susceptible to 2,4-D and MCPA and bentazon has been used successfully against this species in Brazil. Among residual herbicides good results have been reported with such materials as terbutryne, atrazine/alachior mixtures, methabenzthiazuron and pendimethalin. Residual chemicals effective against *R. scabra* include atrazine, metolachlor and trifluralin.

RUBIACEAE

Spermacoce pusilla Wall. (= *Borreria stricta*   Buttonweed
    auct. non (L.f.) G.F. Mey.)
S. princeae (K. Schum.) Verdc. (= *Borreria princeae* K. Schum.)

DESCRIPTION

S. pusilla is an annual, much branched, erect, bushy herb growing to a height of about 30 cm with rough, hairy stems, simple leaves in opposite pairs and small, white flowers with purple veins, clustered in circular heads in the leaf axils and at the ends of short, leafy branches. The roughly hairy leaves are up to 5 cm long × 5 mm wide, with a pointed tip and sessile base. The flower clusters are about 1 cm diameter and borne along the greater part of the stem. The flowers themselves are sessile and 3 mm long, consisting of 4 sepals, a 4-lobed tubular corolla and 4 stamens above a 2-celled ovary. When ripe the fruit divides into 2 one-seeded sections.

There are a number of other *Spermacoce* species in East Africa which may occasionally occur as weeds, but the only other species recorded as being important is *S. princeae,* a perennial with long, weak, sprawling branches. It can be readily distinguished from *S. pusilla* by its habit of growth and by its broader, lanceolate, prominently veined leaves up to 5 cm long × 2.5 cm wide.

Distribution and importance
Both species are indigenous. *S. pusilla* is distributed throughout East Africa and is common over a wide range of altitudes in Tanzania and Uganda, especially on the poorer soils. It appears to be less frequent in Kenya, but is recorded as common in the Eldoret and Nandi districts. It is principally a weed of cultivation and also occurs quite commonly in

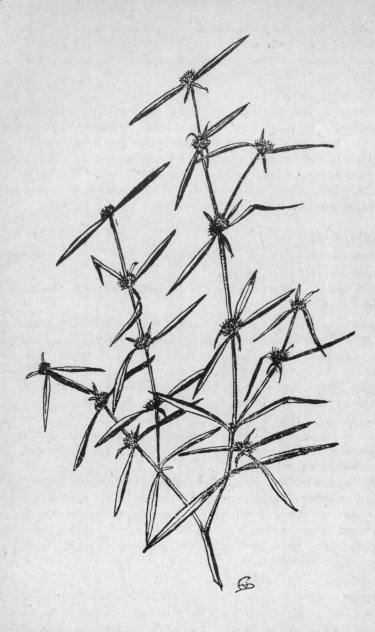

Figure 122. Spermacoce pusilla

poor grassland. *S. princeae* is principally a plant of forest edges and clearings at higher altitudes. It is local, but important, as a weed of high altitude pastures in parts of the Kenya highlands.

Methods of control
S. princeae is little affected by 2,4-D or MCPA. On another perennial species in Puerto Rico good control is reported with dicamba or picloram and in Malaysia glyphosate has been used successfully.

There is no reported information on *S. pusilla* but similar annual species have been found relatively resistant to 2,4-D in the Sudan. Residual herbicides have been found more effective than foliar-applied chemicals on *S. senensis* (= *Borreria scabra*) in Zimbabwe and an atrazine/paraquat combination has been used successfully on another species in Ghana.

SCROPHULARIACEAE

Striga hermonthica (Del.) Benth.	Purple Witchweed
S. asiatica (L.) O. Ktze. (= *S. lutea* Lour.)	Red Witchweed
S. forbesii Benth.	
S. gesnerioides (Willd.) Vatke	Tobacco Witchweed

DESCRIPTION

A group of single-stemmed or branched, annual, parasitic herbs growing from 10 to 60 cm high, with a reduced root system, simple leaves which are opposite below and alternate above, and showy, 2-lipped flowers in terminal spikes.

S. hermonthica reaches a height of up to 1.2 m and has a roughly hairy, 4-angled stem. The leaves are narrowly lanceolate, up to 8 cm long, pointed at the tip and sessile at the base, with a rough surface and sparsely toothed margin. The flowers are reddish purple, 2 —3 cm across and arranged in rather dense spikes about 15 cm long. They are sessile and arise singly in the axils of small, narrow bracts. The calyx is shortly tubular, with 5 long, pointed teeth and is much shorter than the 1 —2 cm corolla tube. The corolla expands above into a shallowly 2-lobed upper lip and a flat lower lip which is deeply divided into 3 rounded lobes, indented at the tip. There are 4 stamens attached inside the tube of the corolla and the fruit is a capsule containing numerous, minute seeds.

S. asiatica is smaller, rarely exceeding 25 cm high, has very narrow leaves up to 4 cm long and the smaller flowers are usually a bright, vermilion red, though they may also be yellow or white. The bracts in the axils of which the flowers arise are similar to the leaves, but smaller, and the spikes may be dense or loose. The corolla tube is very slender and

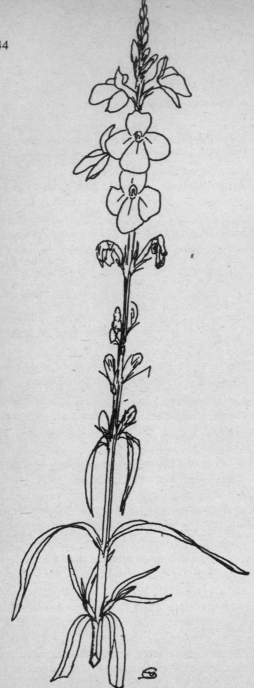

Figure 123. Striga hermonthica

about 1 cm long. Seed dormancy of up to 20 years has been reported.

S. forbesii grows to about the same height as *S. hermonthica* and has similar leaves. It differs in the flowers, which are usually pink, and rather few in number, arising in the axils of the rather widely separated, upper leaves. The corolla tube is about 2.5 cm long.

S. gesnerioides is quite distinctive because of its tuberous roots, purplish stems, scale-like, often purplish leaves not exceeding 12 mm long and rose or pale purple flowers in loose, elongated spikes.

Distribution and importance
All four species are widely distributed on the African continent and are found in many parts of East Africa growing mainly on indigenous grasses. Although of widespread occurrence they are of rather localized importance as weeds and are especially important on poor, sandy soils. *S. hermonthica* is the most important, especially in the Tabora district of Tanzania, around the estern shores of Lake Victoria, both in Tanzania and Kenya, and in the Mbale, Serere and Gulu districts of Uganda. In these areas it causes considerable injury to crops of maize, sorghum, millet and on occasion sugar cane. A single crop plant may be parasitized by a large number of *Striga* plants and the growth checked to such an extent that little gain is produced.

S. asiatica is also common as a weed and parasitizes similar crops, but rarely reaches the importance in East Africa that it does in the Sudan or South Africa. *S. forbesii* is less common, but is reported as a weed in Zanzibar, in grass leys in the Nandi district of Kenya and as a parasite of rice in Tanzania. *S. gesnerioides* although of common occurrence in East Africa is as yet only of potential importance as a weed. In Zimbabwe it is sometimes troublesome as a parasite on tobacco.

Methods of control
Most of the work on control of *Striga* has been done on *S. asiatica* and *S. hermonthica,* but the principles involved are the same for all the species. In the Sudan *S. hermonthica* is worst on poor soils exhausted by continuous cropping, and application of nitrogenous fertilizer has often given larger yield increases than control of the parasite. Crops can also be badly affected on more fertile soils, however, so that control methods are required in addition to adequate fertilizer treatment.

Under suitable conditions successful control can be obtained by the use of trap crops. Various *Sorghum* species, or other grasses parasitized by *Striga* can be sown to stimulate germination of the seed and then ploughed in before the weed reaches maturity. By repeating this process several times in a season the number of seeds in the soil can be greatly reduced. In the Sudan, *Sorghum sudanense* (Sudan grass) has been much used in this way against *S. hermonthica* and can be ploughed in 5 weeks after sowing. Various other crops can be sown which stimulate germination, but are not parasitized or fail to support the weed to maturity and, by growing a succession of such crops for several years, eradication can be accomplished gradually. Groundnut, several other legumes and sunflower have been employed for this purpose, mainly against *S. asiatica,* in South Africa. In West Africa early sowing of crops and attention to preventing seed production are recommended as methods of limiting *Striga* problems.

Herbicides can be used to control *Striga.* 2,4-D or MCPA, for example, will kill established plants, but germination continues after the residues of the chemical have disappeared and the weed exerts much of its harmful effect on the crop before it emerges above ground. A similar limitation applies to the use of soil-applied herbicides which, in general, have only given moderate results. In the Sudan, however, a combination of atrazine and nitrogenous fertilizer has been found promising against *S. hermonthica* in sorghum and gives better results than would be expected from the single effects of the two treatments.

Striga seed will only germinate in the presence of a suitable chemical stimulant. A natural stimulant produced by host plants has been isolated and identified as strigol, and various synthetic analogues (GR7 etc.) have been produced which are also effective stimulants. Field scale use of these chemicals is not yet a practical possibility but, in the USA, successful results in freeing the soil of *S. asiatica* seed have been obtained by injecting ethylene. This gas stimulates germination of the seed and results in subsequent death of the seedlings in the absence of a suitable host plant.

For the majority of situations in East Africa where *Striga* is a problem the most promising control technique is probably the development of resistant crop cultivars and considerable progress is now being made in breeding sorghum cultivars resistant to *Striga* attacks.

SOLANACEAE

Datura stramonium L.   Thorn Apple
D. metel L.

DESCRIPTION

D. stramonium is a stout, erect, annual herb growing 1.2 m or more high, with a stem which forks repeatedly into two, alternate leaves, large, erect, white, funnel-shaped flowers and thorny fruit. The leaves have stalks up to 10 cm long and blades up to 20 cm long × 15 cm wide, with sharply and irregularly lobed or toothed margins, a sharply pointed tip and an abruptly or gradually narrowed base. The flowers are up to 8 cm long and are borne singly in the leaf axils on stalks not exceeding 1.5 cm. They consist of a narrow, 5-angled, tubular calyx 2.5 —4 cm long which terminates in 5 short teeth, a corolla with a long tube and 5 spread-

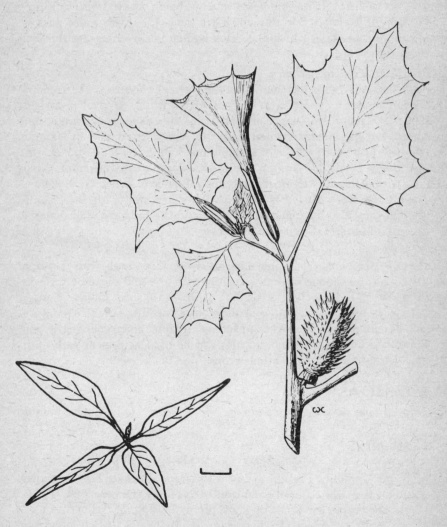

Figure 124 *Datura stramonium*

ing, narrowly pointed lobes, 5 stamens attached to the corolla tube and a 2-celled ovary, developing into a thorny fruit up to 4 cm long × 2.5 cm wide and containing numerous dark brown seeds. (A very similar species, *D. ferox* L., with smaller flowers and fewer, larger spines up to 2.5 cm long occurs in South Africa, but has not yet been recorded in East Africa).

D. metel differs in its youngest branches being covered with soft, grey hairs; in its 10-lobed flowers, which are long (15—18 cm) and narrow, and in its fruit stalk, which elongates and curves over so that the fruit hangs downwards.

Distribution and importance
A weed of worldwide distribution in the warmer countries, *D. stramonium* is common in most parts of East Africa from sea level to at least 2500 m. *D. metel*, another cosmopolitan species, is less common in East Africa, but has been recorded near the Kenya coast and in Uganda. They are principally weeds of arable crops and waste land and *D. stramonium* in particular is often one of the worst weeds in maize. Apart from their competitive effects on crops, all species are very poisonous. Their seeds have caused poisoning when harvested with cereal crops (though they also have medicinal uses) and the foliage is poisonous to cattle if it becomes mixed with fodder.

Methods of control
Datura species can be controlled as young seedlings with 2,4-D or MCPA but resistance increases rapidly with age and dicamba formulations are more effective in dealing with established plants. Contact herbicides are only effective in the young seedling stage. Among residuals a number of chemicals, including atrazine, ametryne, metribuzin, linuron, pendimethalin and alachlor can be used as pre- or early post-emergence treatments in various crops.

SOLANACEAE

Nicandra physalodes (L.) Gaertn.   Chinese Lantern
                                                      Apple of Peru

DESCRIPTION
A much branched, soft-stemmed, annual herb usually growing 1—1.2 m high, but sometimes reaching 2 m with large, alternate leaves, showy, pale bluish or white flowers and fruit enclosed by the enlarged, winged calyx. The leaves are borne on stalks up to 5 cm long, the blades up to 12 cm long × 7 cm wide, narrowing gradually to a blunt point and abruptly narrowed at the base, with an irregularly and deeply toothed

Figure 125. Nicandra physalodes

margin. The pale blue flowers are 2.5 —4 cm across and borne singly in the axils of the upper leaves on stalks about 2 cm long. There is a 5-winged calyx, half the length of the corolla, which is made up of 5 heart-shaped segments each extended into 2 curved, pointed lobes at the base. The corolla is broadly funnel-shaped with 5 rounded lobes and there are 5 stamens attached to the corolla tube. The ovary develops into a yellow berry which is hidden within the enlarged, brown, membranous, persistent calyx and contains numerous small seeds.

Distribution and importance

An introduced plant originating from South America, *Nicandra* is a common weed of crops and waste land in various parts of Africa. It is very common in parts of East Africa, especially northern Tanzania and the Sotik, Uasin Gishu and Kitale districts of Kenya. It has not been recorded from Uganda. It often forms very dense stands and, in localized areas, is not infrequently the dominant weed. Maize is the crop most commonly infested, but it also occurs in a wide range of other crops. In Zimbabwe *Nicandra* is a host of the eelworm *Meloidogyne javanica*.

Methods of control

In most parts of the world where it occurs as a weed *Nicandra* is listed as susceptible to 2,4-D and MCPA. In some East African trials, however, normal doses have only given effective control when applied in the very young seedling stage or as pre-emergence treatments. In New Zealand good kills have been obtained with MCPB, and in East Africa, 2,4-DB has also proved effective. Good kills of young plants can be obtained with bentazon and pre- or early post-emergence treatment with residual herbicides of the triazine or substitued urea groups also give excellent control.

SOLANACEAE

Physalis ixocarpa Hornem.	Tomatillo, Purple Gooseberry
P. angulata L.	Wild Cape Gooseberry
P. peruviana L.	Cape Gooseberry

P. micrantha Link. (= *P. divaricata* D. Don and often incorrectly called *P. minima* L.)

DESCRIPTION

These four species of *Physalis* are branched herbs growing 0.6 —1 m high with simple, alternate leaves, pale yellow flowers and characteristic, large, inflated calyces enclosing the berries.

    P. ixocarpa is an annual, with leaves on stalks about 2.5 cm long. The ovate blades are up to 6 cm long × 3 cm wide, with a pointed tip, wedge-

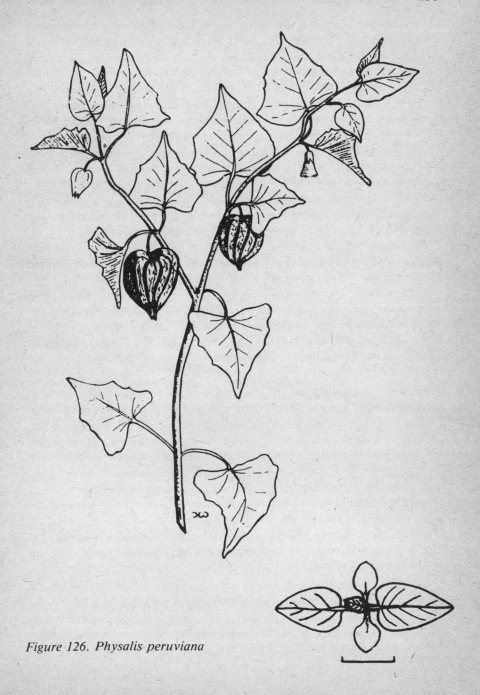

Figure 126. Physalis peruviana

shaped base and entire or shallowly wavy margin. The flowers are borne singly in the upper leaf axils on stalks up to 5 mm long and tend to hang downwards. They are about 1 cm long and 1 cm across with a dark centre. The calyx has 5 teeth about as long as wide and is at first about half as long as the shallowly 5-lobed corolla. There are 5 stamens attached to the corolla tube and the ovary develops into a sticky, purplish berry, containing numerous seeds. As the fruit develops the calyx tube enlarges to form a more or less spherical, papery bladder 2.5 cm or more across.

The annual *P. angulata* is similar in general appearance to the above. The leaves are somewhat larger, up to 10 cm × 5 cm with a more deeply indented margin, and the flower stalks are longer, reaching 12 mm or more. The flowers themselves are smaller, about 5 mm long, the calyx teeth are distinctly longer than broad, while the fruit is yellowish, not sticky and enclosed by an ovoid fruiting calyx up to 3 × 2 cm.

P. micrantha is also annual and has leaves about 2.5 cm long and more or less entire margins. The flowers are somewhat smaller than in *P. angulata* and the fruiting calyx, up to 1.5 cm long, much smaller.

P. peruviana is a perennial and the whole plant is softly hairy. The upper leaves are often oppositely arranged, up to 10 cm long × 5 cm wide, and more or less triangular in shape, with entire margins or a few large, shallow teeth, and petioles up to 6 cm long. The flowers are about 1.5 cm across. The inflated calyx is similar in shape to *P. angulata* and the berry is golden when ripe.

Distribution and importance

All four species are native to Central or South America and have been introduced into East Africa. The fruits are edible and *P. peruviana* especially is widely cultivated, often becoming naturalized. *P. ixocarpa* presents the most important weed problem and is one of the more troublesome arable weeds in parts of the Kenya highlands, where it is reported to be spreading rapidly. It does not yet appear to have become established in Tanzania or Uganda. *P. peruviana* and *P. micrantha* are more widely distributed, occurring in all three countries, but are not very serious as weeds. *P. angulata* has been recorded occasionally in cereal crops in Kenya and Tanzania but is of localized importance. In Zimbabwe and South Africa it is common and troublesome in maize.

Methods of control

P. ixocarpa has been found relatively resistant to 2,4-D and MCPA in Kenya and *P. angulata* is recorded as resistant to 2,4-D in Zimbabwe. Combinations of mecoprop with ioxynil and bromoxynil are more effective and can be used successfully against seedlings in cereal crops.

In the USA good results against *P. angulata* in various crops are reported with several residual chemicals, such as atrazine, metribuzin, diuron, linuron and alachlor. The reaction to herbicides of seedlings of the perennial *P. peruviana* is much the same as that of the annual species. Established plants are more difficult to kill but non-selective control with amitrole or glyphosate appears to be possible.

SOLANACEAE

Solanum nigrum L.   Black Nightshade
 (including *S. grossedentatum* A. Rich.)

DESCRIPTION
A branched, annual herb growing to 60 cm high with alternate, simple, rather dark green leaves, small clusters of white flowers with yellow centres and green berries, which usually turn black. Orange and red berried forms also occur. The leaves are ovate, up to 6 cm long × 4 cm wide and pointed at the tip, with the margin wavy or indented into deeper, round teeth and a wedge-shaped base, narrowed gradually into the 1 —2 cm stalk. The clusters of flowers arise from the stems between leaves, not from the leaf axils, and consist of a common stalk with a small number (usually 3 —6) of stalked flowers arising from its apex. The flowers are about 5 mm across and consist of a calyx with 5 blunt teeth, a corolla with 5 lobes which are at first spreading, later curved backwards, and 5 stamens forming a yellow projection at the mouth of the corolla tube. The ovary develops into a many seeded, fleshy, black fruit 5 mm or more in diameter and, as the fruit ripens, the stalks bend downwards.

Distribution and importance
A cosmopolitan weed, found in most tropical and temperate countries and widespread throughout East Africa from sea level to at least 2000 m. It is primarily a weed of arable and waste land and infests a wide range of crops. Apart from its competitive effect on crops *S. nigrum* is also a poisonous plant, though its toxicity appears to be variable. In Zimbabwe, for example, the leaves are reported to be eaten as a vegetable and the ripe fruits are also edible. On the other hand, unripe fruits are poisonous, sometimes only slightly, at other times very much so, especially to children, and cattle have also been affected by grazing on the foliage.

Methods of control
S. nigrum is moderately susceptible to 2,4-D and MCPA but the plants are only satisfactorily killed as small seedlings. Mixtures containing bromoxynil, ioxynil or dicamba are more effective. Contact herbicides including bentazon give good control and various residual herbicides are

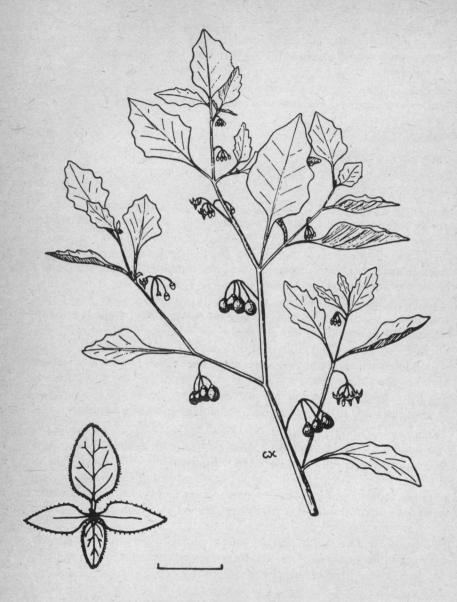

Figure 127. Solanum nigrum

herbicides are effective as pre- or early post-emergence treatments, including atrazine, diuron, linuron, alachlor and pendimethalin. Metribuzin, however, is not effective against *Solanum* species.

SOLANACEAE

Solanum incanum L.   Sodom Apple
 (including *S. panduriforme* Dunal, *S. bojeri* Dunal,
 S. campylacanthum A. Rich., *S. delagoense* Dunal and
 S. obliquum Dammer)

DESCRIPTION

S. incanum is a shrubby, perennial, thicket-forming herb, with branched, prickly stems up to 1.5 m or more high, growing from an extensive underground rhizome system, and with alternate, simple leaves, purple flowers and fruits like small, yellow, leathery tomatoes. The leaves have short stalks, are lanceolate in shape, up to 12 cm long × 6 cm wide, pointed at the tip, rather abruptly narrowed at the base, and have a shallowly wavy margin. The lower surface is densely covered with grey hairs, and the midrib often bears a few prickles. The flowers are 2 cm or more across and are borne singly, or in groups of up to 5, along short branches arising from the stems between leaf bases and usually directed downwards. They consist of a 5-lobed, often spiny calyx, a corolla with 5 spreading, somewhat reflexed lobes, 5 stamens forming a conical projection around the style and an ovary developing into a many seeded, globular berry 2 cm or so in diameter.

Distribution and importance
S. incanum is the common East African Sodom Apple, occurring on waste ground and roadsides, readily invading all but the most vigorous pastures and frequently developing into dense thickets which are avoided by cattle. It is noted as a weed in most parts of Kenya, Tanzania and Uganda, and is recorded as an important problem from more areas than any other weed dealt with in the survey. In Tanzania the fruits are said to be eaten in times of famine. In Zimbabwe they are regarded as poisonous when unripe.

Methods of control
Mowing appears to encourage Sodom Apple rather than control it, while controlled grazing and burning also seem to be relatively ineffective. It can be readily controlled with 2,4-D, however, which has been shown to be appreciably more effective than MCPA. Either the amine or ester formulation of 2,4-D can be employed and the best results are obtained by spraying when the majority of the weed is in the early flowering stage.

Figure 128. Solanum incanum

A single application gives a high percentage kill. Some regrowth usually takes place, but two applications, either during the same rainy reason it it is long enough or in successive seasons, give a large measure of eradication.

At Kongwa in Tanzania 2,4-D has been employed against Sodom Apple on a large scale and it has been found advantageous to burn or cut the weed about a month before the rains are due to begin. It is then possible to apply one spray early in the rainy season and another 2—3 months later, so obtaining almost complete control in one season.

TILIACEAE

Triumfetta rhomboidea Jacq.
T. flavescens A. Rich.

DESCRIPTION

T. rhomboidea is an annual (or sometimes perennial), shrubby herb growing 1—1.2 m or occasionally up to 2 m high, with alternate leaves, axillary clusters of small, yellow flowers and small, prickly fruit. The leaves are borne on stalks up to 5 cm long. The blades are ovate, up to 15 cm long × 10 cm wide and pointed at the tip, usually with 2 lateral, acutely tipped and forwardly directed lobes. The leaf base is rounded, there is an irregularly toothed margin and 3—7 prominent nerves diverge from the junction with the stalk. The flowering part of the stem is leafy, the leaves being narrower than those lower down. The flower clusters have very short stalks. The individual flowers, which are about 8 mm across, are also shortly stalked. They consist of 5 narrow sepals, slightly longer than the 5 narrow petals, about 15 stamens and an ovary of 2—3 segments, developing into a hairy, more or less spherical fruit, 5 mm across and covered with hooked prickles.

T. flavescens is a larger, shrubby perennial, with hairy, black-spotted stems. The leaves are almost as broad as long, only slightly 3-lobed and densely hairy on both surfaces. The flowers are similar to those of *T. rhomboidea*, but the petals are shorter, about half the length of the sepals, there are 20—22 stamens and the fruit is ovoid, with the prickles curved inwards rather than projecting outwards.

Distribution and importance

T. rhomboidea occurs throughout the tropics and is common and widespread in East Africa. It occurs mainly as a weed of waste ground and grassland, less often as an arable weed. *T. flavescens* is more restricted in its distribution to higher altitudes in Kenya, Uganda and north-east Tanzania, where it is a weed of grassland on cleared forest areas. It pre-

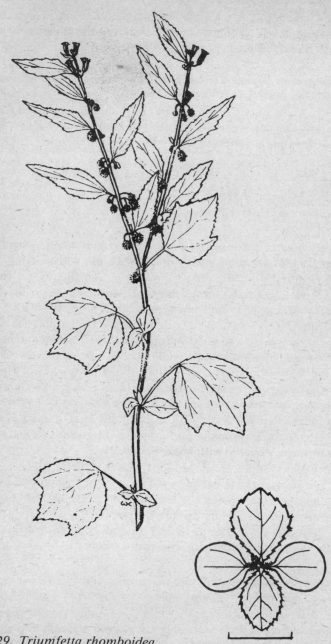

Figure 129. Triumfetta rhomboidea

sents a particular problem at Marsabit, in northern Kenya, where it has invaded a large area of grazing land following protection from burning.

Methods of control

In Zimbabwe *T. rhomboidea* is reported susceptible to pre-emergence treatment with 2,4-D, but relatively resistant to post-emergence treatment. There is no information on its control in East Africa, nor on the effects of chemicals on *T. flavescens*. The latter species can be kept in check by occasional, controlled grass-burning and does not appear to be particularly difficult to remove mechanically or by cultivation. Directed spraying or wiper-bar application of glyphosate would be expected to control these species.

URTICACEAE

Urtica massaica Mildbr.   Stinging Nettle

DESCRIPTION

A perennial herb covered with stiff, fiercely stinging hairs and with erect, simple branches up to 2 m or more high, growing from a branched, creeping stem, situated at ground level or just below and rooting at the nodes. The opposite leaves are borne on stalks up to 10 cm long and have heart-shaped blades up to 15 cm long × 10 cm wide, narrowed gradually to a pointed tip and with sharply and deeply toothed margins. The flowers are small and green, very numerous and borne in clusters along the branches of axillary inflorescences, which are up to 6 cm long and hang downwards. The male flowers have 4 stamens, the female an ovary with numerous styles, which develops into a single-seeded fruit.

Distribution and importance

An indigenous plant of wetter areas and altitudes above 1500 m, growing mainly on well-drained fertile soils and at forest edges. It is recorded as a weed from various parts of the Kenya highlands, especially in the Elgeyo-Marakwet and Naivasha districts and on the Kinangop. In Tanzania it is found in the Usambara, Southern Highlands and Arusha districts and, in Uganda, occurs in Kigezi. It is of importance principally as a weed of upland grassland and forestry plantations, but sometimes occurs also in arable crops.

Methods of control

There is only a limited amount of information on the effects of herbicides. MCPA has been found to give good control of the aerial parts but regrowth generally occurs. Formulations containing dicamba, picloram

Figure 130. Urtica massaica

or clopyralid would be expected to give more permanent control and gly phosate has proved effective against the related *U. dioica* in Europe.

ZYGOPHYLLACEAE

Tribulus terrestris L.   Caltrops, Puncture Vine
T. cistoides L.

DESCRIPTION
Taprooted, annual herbs, with hairy, prostrate radiating branches up to 1 m long, pinnate leaves arranged in opposite pairs, yellow flowers, which open only in the morning, and 5-angled spiny fruits. One of each leaf pair is smaller than the other, and large and small leaves alternate along the stem. The leaves are up to 8 cm long and consist of 6 —8 pairs of unstalked, lanceolate leaflets up to 12 mm long × 3 mm wide, borne along a central midrib. The flowers of *T. terrestris* are about 1 cm across and are borne singly in the axils of the smaller leaves on 1 —2 cm stalks. They consist of 5 sepals and petals, 10 stamens and a 5-celled ovary. The fruit splits into 5 sections when ripe, the sections being more or less triangular in shape, with numerous small and 2 large, very sharp spines projecting at the tip. The two species are similar in appearance, but the flowers of *T. cistoides* are much larger and generally 4 —5 cm in diameter.

Distribution
T. terrestris is a cosmopolitan weed, occurring in most parts of the tropics and sub-tropics and extending into warmer temperate regions. It is widely distributed in East Africa, mostly at lower altitudes, but reaches at least 1500 m in drier areas. In Kenya it is reported as especially common in the Machakos district. It occurs in maize and other arable crops, in poor grassland and in waste ground. The spiny fruits can be troublesome in the feet of man and animals, and in South Africa it often causes poisoning when grazed by sheep. *T. cistoides* is also common but less often recorded as a weed.

Methods of control
Post-emergence treatment with 2,4-D or MCPA gives control in the young seedling stage but resistance increases rapidly with age. Among contact herbicides bentazon has given good control of young seedlings in groundnuts in Israel. Among the residuals prometryne and linuron are reported to be effective and dinitramine has been used successfully as a

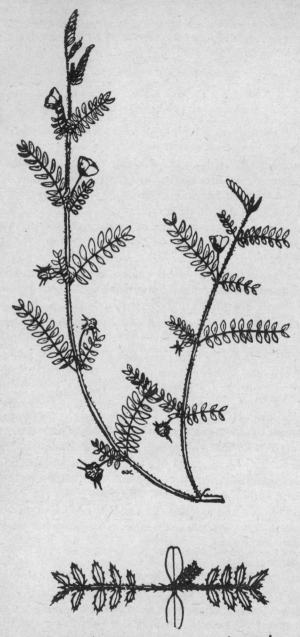

Figure 131. Tribulus terrestris

soil-incorporated treatment. Considerable success has been obtained in the USA with biological control, the insects employed being weevils of the genus *Microlarinus*.

List of Poisonous Plants

Ageratum conyzoides	— foliage, suspect	120
Anagallis arvensis	— foliage	236
Argemone mexicana	— seed	211
Centaurea melitensis	— foliage, suspect	127
Chenopodium ambrosioides	— foliage, suspect	112
Crotalaria incana subsp.		
purpurascens	— foliage	213
Crotalaria polysperma	— foliage, at times	213
Cynodon dactylon	— foliage, at times	31
Cynodon nlemfuensis	— foliage, at times	31
Datura spp.	— seed and foliage	246
Emex australis	— foliage	221
Indigofera spicata	— foliage	216
Lantana camara	— foliage and berries	82
Leonotis spp.	— foliage, suspect	185
Lolium temulentum	— seed	46
Oxalis corniculata	— foliage	207
Phytolacca dodecandra	— foliage	219
Pistia stratiotes	— foliage, suspect	1
Pteridium aquilinum	— foliage	232
Ricinus communis	— seed	183
Rumex acetosella	— foliage, suspect	229
Senecio discifolius	— foliage, suspect	151
Senecio ruwenzoriensis	— foliage	153
Senecio moorei	— foliage, suspect	153
Sida rhombifolia	— fruit	200
Solanum incanum	— berries, at times	255
Solanum nigrum	— foliage and berries, at times	253
Tribulus cistoides	— foliage	261
Tribulus terrestris	— foliage	261

(For a full treatment consult Watt & Breyer-Brandwijk, 1962, and Verdcourt & Trump, 1969.)

Bibliography

The following are the principal works which have been consulted.

AGNEW, A. D. Q. *Upland Kenya Wild Flowers.* Oxford University Press, London, 1974.

BLUNDELL, M. The Wild Flowers of Kenya. Collins, London, 1982.

BOGDAN, A. V., *A Revised List of Kenya Grasses.* Government Printer, Nairobi, 1958.

BOGDAN, A. V. and PRATT, D. J., *Common Acacias of Kenya.* Government Printer, Nairobi, 1961.

BRENAN, J.P.M. AND GREENWAY, P.J., *Tanganyika Check List, Part 2.* Imperial Forestry Institute, Oxford, 1949.

DALE, I. R. and GREENWAY, P. J. *Kenya Trees and Shrubs.* Buchanans Kenya Estates, Nairobi, 1962.

EDWARDS, D. C. and BOGDAN, A. W., *Important Grassland Plants of Kenya.* Pitman, London, 1951.

EGGELING, W. J. and DALE, I. R., *Indigenous Trees of Uganda* (Second Edition). Government Printer. Entebbe, 1952.

FRYER, J. and MAKEPEACE, R. (Editors). *Weed Control Handbook Vol II Recommendations* (7th edition). Blackwell Scientific Publications, Oxford, 1972.

HARKER, K. W. and NAPPER, D., *An Illustrated Guide to the Grasses of Uganda.* Government Printer, Entebbe, 1960.

HOCOMBE, S. D. and YATES, R. J., *Guide to Chemical Weed Control in East African Crops.* East African Literature Bureau, Dar es Salaam, 1963.

LIND, E. M. and TALLANTIRE, A. C., *Some Common Flowering Plants of Uganda.* Oxford University Press, London, 1962.

NAPPER, D., *Grasses of Tanganyika.* Government Printer, Dar es Salaam, 1966.

Proceedings of the First East African Herbicide Conference, Nairobi, 1957.

Proceedings of the African Weed Control Conference, Victoria Falls Hotel, 23 – 25 July 1958.

Proceedings of the Third East African Herbicide Conference, Nairobi, 1964.

Proceedings of the 4th East African Herbicide Conference, Arusha, 1970.

Proceedings of the 5th East African Weed Control Conference, Nairobi, 1974.

Proceedings of the 6th East African Weed Science Conference, Nairobi, 1976.

Proceedings of the 7th East African Weed Science Conference, Nairobi, 1979.

Proceedings of the 8th East African Weed Science Society Conference, Nairobi, 1981.

TERRY, P. J. Sedge Weeds of East Africa. I. Identification. East African Agriculture and Forestry Journal, Vol. 42, 1976, pp. 231—249.

TERRY, P. J. Sedge Weeds of East Africa. II Distribution. Weed Research Organization Technical Report No. 50. Oxford, UK, 1978.

TILEY, G.E.D. Weeds. In *Agriculture in Uganda* (Edited J.D. Jameson). Oxford University Press, London, 1970. pp. 297—317.

TURRILL, W. B. and MILNE-REDHEAD, E. (Editors), *Flora of Tropical East Africa*. H.M. Stationery Office, London. Government Printers, Nairobi, Dar es Salaam and Entebbe.

Capparaceae, 1964.
Caryophyllaceae, 1956.
Chenopodiaceae, 1954.
Convolvulaceae, 1963.
Cruciferae, 1982.
Geraniaceae, 1971.
Gramineae (Part 1), 1970.
Gramineae (Part 2), 1974.
Gramineae (Part 3), 1982.

Leguminosae, Caesalpinioideae, 1967.
Leguminosae, Mimosoideae, 1959
Leguminosae, Papilionoideae, 1971.
Orobanchaceae, 1957.
Oxalidaceae, 1971.
Papaveraceae, 1962.
Phytolaccaceae, 1971.
Polygonaceae, 1958.
Pontederiaceae, 1968.
Primulaceae, 1958.
Rubiaceae (Part 1), 1976.
Typhaceae, 1971.

VERDCOURT, B. and TRUMP, E. C. *Common Poisonous Plants of East Africa.* Collins, London, 1969.

WATT, J. M. and BREYER-BRANDWIJK, M. G. *The medicinal and poisonous plants of Southern and Eastern Africa*. E. & S. Livingstone, Edinburgh & London, 1962.

WILD, H. *Common Rhodesian Weeds*. Government Printer, Salisbury, 1958.

WILD, H. *Harmful Aquatic Plants in Africa and Madagascar.* CCTA/ CSA, Salisbury, 1961.

WILLIAMS, R. O. *Useful and Ornamental Plants of Zanzibar and Pemba.* Government Printer, Zanzibar, 1949.

Glossary

alternate, arising singly from the stem or branches at different levels.

amine, amine salt, a type of water-soluble formulation of acid herbicides, especially 2,4-D and related compounds.

annual, a plant completing its life cycle in less than one year.

annular, ring-shaped.

anther, the part of the stamen which produces pollen.

antidote, a chemical additive which, when mixed with certain herbicides (particularly thiocarbamates), reduces crop injury without affecting the activity against weeds; i.e. increases selectivity.

aromatic, with a strong, spicy scent.

auricles, (of grasses) a pair of small, pointed projections at the base of the leaf blade, often clasping the stem.

auxin, a natural, plant growth-regulating chemical.

auxin-type growth-regulator (synthetic growth-regulator), a herbicide causing somewhat similar effects on plants to naturally occurring auxins.

axil, the angle between a leaf and the branch from which it arises.

axillary, arising in the leaf axil.

axis, the main stem of a plant or an inflorescence.

basal-bark, a type of herbicide application used on woody plants where the chemical is applied to the intact or cut bark at the base of the trunk.

beak, a pointed projection at the tip of a fruit or seed.

biological control, the use of living organisms to control pests. The organisms used for biological control of weeds are most commonly insects.

blade, the expanded portion of a leaf.

bract, a more or less modified leaf at the base of a flower or flower head.

bulb, an underground storage organ made up of fleshy leaf-bases.

bulbil, a small bulb.

calyx, the outer ring of flower structures, made up of separate or joined sepals.

capsule, a dry fruit which opens at maturity to release the seeds.

carpel, one of the innermost ring of flower structures containing an ovary and, later, seeds.

claw, the narrow, basal part of a petal.

clover-like (of leaves), made up of three leaflets borne at the end of a relatively long stalk.

composite, the type of flower found in the Compositae. The apparent flower consists of a number of small flowers aggregated into a head and surrounded by one or more rows of bracts.

compound (of leaves), made up of several distinct leaflets.

contact, a type of herbicide which affects the leaf tissues with which it comes in contact, but is not moved within the plant.

contiguous, adjoining.

coppice, shoots of a woody plant growing from the base of the trunk after cutting.

corolla, the ring of usually showy flower structures within the calyx made up of separate or joined petals.

corolla tube, the tube formed by the joined bases of the petals.

digitate (of leaves), a compound leaf with several leaflets diverging from the top of the stalk.

diploid, the normal type of plant with two similar sets of chromosomes in the nuclei of its cells.

direct drilling—see *minimum tillage*

directed, a type of application in which herbicide is directed so as to cover the weeds but avoid the crop plants.

disc-floret, one of the central florets of a composite flower.

elliptic (of leaves), broadest about the middle and rounded at each end.

emulsifiable oil, a type of herbicide formulation in which the active material is dissolved in oil and emulsifiers added so that it can be mixed with water.

entire (of leaves), the margin without teeth or lobes.

ester, a type of formulation of acid herbicides (especially the synthetic growth-regulators) which is soluble in oil but requires the addition of an emulsifier to permit mixing with water.

female flower, a flower with functional ovaries but no stamens.

fertile (of flowers), capable of producing viable seed.

filament, the stalk of a stamen.

floret, one of a group of small flowers making up a flower head.

flower head, a group of flowers clustered into a head.

foliar-acting herbicides, a herbicide which exerts most of its effect via the leaves. It may or may not also have an effect via the roots.

formulation (of pesticides), the way in which chemical compounds are prepared for practical use.

frill, a ring of overlapping cuts in the bark of a free made to assist penetration of herbicide.

frond, the leaf of ferns which bears spore-producing structures.

fruit, the dry or fleshy structure containing the ripe seeds.

fused, the parts joined together.

glabrous, without hairs.

gland, a small globular structure containing liquid, either sunk into the leaf or borne at the tip of a slender stalk (glandular hair); a small, fleshy protuberance.

glume, see *spikelet*

granular formulations, a formulation prepared for dry application in which the active chemical is incorporated in or on granules of an inert carrier, often clay.

herb, a non-woody plant.

incorporation (of chemical), mixing shallowly into the soil. Usually necessary with volatile chemicals to prevent loss into the atmosphere.

indigenous, native.

inflorescence, the flowering part of a plant or branch.

internode, the section of stem between two nodes.

interrupted, not continuous.

involucre, a ring of bracts surrounding a flower head, often with the appearance of a calyx.

keel, the two partially joined petals of papilionaceous flowers which form a sharp edge resembling the keel of a boat; any sharp edge which runs longitudinally.

lanceolate (of leaves), shaped like the head of a spear, with a pointed tip and the broadest part towards the base.

latex, milky juice.

leaflet, one of the component parts of a compound leaf.

leaf scar, the scar left on a stem after a leaf has fallen.

ligule (of grasses), a short, membranous flap of tissue pressed close to the stem at the junction of the leaf blade and its sheath.

linear, long, narrow and more or less parallel edged.

lip, one of the lobes of a fused corolla which is partially split into two segments as in many Labiatae.

lobed (of leaves), partially divided but not deeply enough to form separate leaflets.

male flower, a flower with functional stamens but no ovary.

membranous, thin and transparent but not green.

midrib, the main vein.

minimum tillage, a system of growing crops with the minimum of soil disturbance, using herbicides (commonly paraquat or glyphosate) to kill existing vegetation at the time of (or prior to) planting. Tillage may either be reduced or, in the case of zero-tillage, dispensed with. Direct drilling is another term used to describe this system.

nerve, one of the main veins of a leaf or flower part.

node, the point on a main stem or branch where leaves or buds arise.

nutlet, a small nut, used with special reference to the 4 one-seeded fruits of Boraginaceae.

oblong (of leaves), longer than broad with roughly parallel sides.

oblique (of leaves), when the two sides of the blade are unequal at the base.

oblong (of leaves), longer than broad with roughly parallel sides.

oblanceolate (of leaves), similar to lanceolate but with the broadest part nearer the tip.

obovate (of leaves), egg-shaped, with the broadest part nearer the tip.

opposite, arising in pairs on opposite sides of the stem or branches.

ovary, the lower part of the carpel which contains the ovules and eventually forms the fruit.

ovate, oval, egg-shaped with the broadest part nearer the base.

overall, a type of application in which herbicide is applied over the whole area, covering both weeds and crop, or over the foliage of a woody plant rather than to the trunk alone.

ovoid, an egg-shaped solid.

ovule, the young seed produced in the ovary before fertilization.

palmate (of leaves), divided into lobes or leaflets like the palm of a hand.

panicle, a loosely branched inflorescence.

pappus, a ring of hairs or scales round the top of the fruit of Compositae.

parasite, a plant often lacking chlorophyll, which obtains its food and water supplies from another living plant.

pea-flower, a type of flower found in the Papilionaceae in which the 5 petals are arranged to form an upright standard, two lateral wings and a keel formed by the fusion of the two lower petals.

pedicel, the stalk of a flower.

peduncle, the common stalk of a group of flowers.

perennial, a plant living for two or more years.

perianth, the outer rings of flower parts, including both sepals and petals.

petal, one of the inner, usually showy, perianth segments.

petiole, the stalk of a leaf.

pinna, one of the primary divisions of a pinnate leaf. The pinnae may themselves be pinnate.

pinnate, a compound leaf with the leaflets arranged on opposite sides of a common stalk.

pod, a type of fruit typical of the Papilionaceae which splits longitudinally in two valves.

post-emergence, a type of herbicide application in which the chemical is applied after the emergence of the crop. Sometimes, but less correctly, used to refer to applications made after weed emergence.

potassium salt, a type of water-soluble formulation of acid herbicides, especially MCPA, mecoprop, dichlorprop and other synthetic growth-regulators.

pre-emergence, a type of herbicide application in which the chemical is applied before the emergence of the crop. Sometimes, but less correctly, used to refer to applications made prior to weed emergence.

pre-planting, a type of herbicide application in which the chemical is applied before the crop is sown or planted. Generally necessary with chemicals which have to be incorporated in the soil.

procumbent, lying loosely on the ground.

prostrate, lying closely on the ground.

raceme, a more or less conical inflorescence with flowers arising laterally from a common stalk, the youngest towards the tip.

ray floret, one of the outer florets of a composite flower when distinct from the disc florets.

receptacle, the flattened or rounded terminal portion of the stem bearing the various parts of the flower of florets.

reflexed, bent sharply backwards.

residual, a type of herbicide application in which the chemical is applied to clean soil and prevents the establishment of weeds as long as it persists.

rhizome, an underground, more or less thickened, usually perennial stem. adj. *rhizomatous.*

rhombic, rhomboid (of leaves), diamond-shaped.

root-crown, the part of a multi-stemmed perennial plant lying just below the soil surface which gives rise to roots and shoots.

rope-wick applicator, see *wiper-bar*

runner, an elongated stem growing horizontally above ground and rooting at the nodes to form new plants.

safener, another term for *antidote.*

scale, a thin, dry structure, not green; often a reduced leaf.

senescent, becoming over-mature; as leaves senesce penetration of herbicide is reduced and less translocation occurs.

sepal, one of the outer, usually green, perianth segments.

serrated, a leaf margin with regular, pointed teeth.

sessile, without a stalk.

sheath, the lower part of a leaf enclosing the stem. A particular feature of grasses.

shrub, a woody plant with spreading branches arising from near the base.

simple (of leaves), not divided into leaflets.

sodium salt, a type of water-soluble formulation of acid herbicides, especially MCPA, MCPB and other synthetic growth-regulators.

soil-applied, a type of herbicide which gains entry into plants predominantly through the roots, but has little effect on foliage. Many residual herbicides are of this type.

spathe, a large bract partially or wholly enclosing a group of flowers.

spike, a raceme of sessile flowers.

spikelet, a small spike of one or more flowers surrounded by bracts. As applied to grasses the term refers to groups of one or more florets subtended by a pair of bracts known as glumes.

spore, a small, unicellular reproductive unit produced in large numbers by ferns and other non-flowering plants.

spot treatment, application of herbicide to individual weeds or patches of weed rather than to whole area.

stamen, the male part of a flower producing pollen.

sterile (of flowers), not capable of producing viable seed; (of shoots) non-flowering.

stigma, the uppermost part of the ovary to which pollen grains adhere.

stipule, leaf or scale-like structures arising from the junction of leaf stalk and stem.

stolon, an elongated stem generally growing along the ground and rooting at the nodes.

strap-shaped, a term applied to the corollas of the ray florets of many Compositae which are several times as long as broad, more or less parallel-sided and square-cut at the tip with 3 or 5 teeth.

style, the upper part of the ovary terminating in the stigma.

subtend, to be situated at the base of (used of bracts at the base of flowers, etc.)

sucker, a shoot arising from the roots of a woody plant, often some distance away from the main stem.

tap root, an unbranched, vertically descending root.

tepal, a segment of a perianth which is not differentiated into sepals and petals.

terminal, borne at the end of a stem.

tetraploid, a type of plant with four similar sets of chromosomes in the nuclei of its cells and usually more robust than a corresponding diploid.

thallus, an undifferentiated plant body.

tiller, a side-shoot arising from the base of a grass plant.

translocated, a type of herbicide which, when applied to the leaves, penetrates into the tissues and is moved to other parts of the plant.

tree, a large woody plant with a single trunk.

trifoliate (of leaves), made up of three leaflets.

tube, the lower part of a fused calyx or corolla.

tuber, a short thickened portion of underground stem bearing dormant buds. adj. *tuberous.*

umbel, a type of inflorescence in which a number of divergent pedicels arise from the top of the stem.

unisexual, flowers bearing either functional stamens or ovaries but not both.

valve, one of the sections of a capsule which has split open.

volume rate, the total amount of spray liquid applied per unit area:
  a high volume spray represents more than 600 l/ha
  medium volume spray represents more than 200 —600 l/ha
  low volume spray represents more than 50 —200 l/ha
  very low volume spray represents more than 5 —50 l/ha
  ultra low volume spray represents less than 5 l/ha.

whorl, an arrangement of leaves or flowers in a circle.

wing, a flattened extension of a stem or leaf stalk.

wiper-bar, a herbicide applicator which is moved over the ground at a certain height so that it makes contact with taller growing weeds and wipes them with small volumes of concentrated chemical (usually by means of a rope-wick) without contacting the lower growing crop, thus providing selective application.

List of Vernacular Names

Akaganda	(Ankole)	*Acacia hockii*
Akanoko	(Ankole)	*Launaea cornuta*
Akasandasanda	(Luganda)	*Euphorbia hirta*
Akatooma	(Ankole)	*Gutenbergia cordifolia*
Akayobyo akasajja	(Luganda)	*Cleome monophylla*
Amaduudu	(Luganda)	*Datura stramonium*
Areri gome dimtu	(Galla)	*Cymbopogon caesius*
Atilili	(Luo)	*Psiadia punctulata*
Auch-auch	(Luo)	*Achyranthes aspera*
Awayo	(Luo)	*Oxygonum sinuatum*
Bek	(Kipsigis)	*Eleusine indica*
Bhangi	(Shambaa)	*Tagetes minuta*
Bibimbet	(Kipsigis)	*Imperata cylindrica*
Birirwet	(Kipsigis)	*Pteridium aquilinum*
Burburetyek	(Kipsigis)	*Cyperus rotundus*
Butabuta	(Ankole)	*Ageratum conyzoides*
Bwara	(Ankole)	*Acacia brevispica*
Chemenet	(Kipsigis)	*Sphaeranthus suaveolens*
Chemogong'it-cheptitet	(Kipsigis)	*Nicandra physalodes*
Chemorut	(Kipsigis)	*Cynodon nlemfuensis,*
		Digitaria scalarum
Chemosibit	(Kipsigis)	*Leonotis nepetifolia*
Chemul cheberemet	(Kipsigis)	*Silene gallica*
Chemul keldaiwet	(Kipsigis)	*Polygonum convolvulus*
Chemutwet ab koik	(Kipsigis)	*Spergula arvensis*
Chepkumiat	(Kipsigis)	*Psiadia punctulata*
Chepnusiat	(Kipsigis)	*Alternanthera pungens*
Chesaleit ne tendea	(Kipsigis)	*Galium spurium*
Chesitet	(Kipsigis)	*Acacia drepanolobium*
Chikoko	(Chagga)	*Pennisetum clandestinum*
Chonge	(Kikuyu)	*Oxygonum sinuatum,*
		Polygonum aviculare

Danga danga	(Bondei, Pare, Shambaa)	*Portulaca oleracea*
Djadja	(Swahili)	*Commelina benghalensis*
Ebunei	(Karamojong)	*Cyathula orthacantha*
Eileili	(Ateso)	*Trichodesma zeylanicum*
Eiyimiyiem	(Masai)	*Indigofera spicata*
Ejai	(Ankole)	*Tagetes minuta*
Ekibebia	(Ankole)	*Setaria verticillata*
Ekifumufumu	(Luganda)	*Leonotis nepetifolia*
Ekifuula	(Luganda)	*Abutilon mauritianum*
Ekijembajembe	(Ankole)	*Argemone mexicana*
Eluai	(Masai)	*Acacia drepanolobium*
Emachuku	(Chagga)	*Sphaeranthus suaveolens*
Emindie	(Chagga)	*Cyathula cylindrica*
Emitwe y'abagurusi	(Ankole)	*Sphaeranthus suaveolens*
Emongulai	(Masai)	*Triumfetta flavescens*
Emoto	(Ateso)	*Striga hermonthica*
Emuruwa	(Masai	*Cynodon nlemfuensis*
Enaponombenek	(Masai)	*Launaea cornuta*
Enchani-rongai	(Masai)	*Chenopodium opulifolium*
Endiati	(Masai)	*Chenopodium* spp., *Gutenbergia cordifolia, Schkuhria pinnata*
Enguruma	(Masai)	*Eleusine indica*
Enkojet-naju	(Masai)	*Rhynchelytrum repens*
Entengotengo	(Luganda)	*Solanum incanum*
Enyaru	(Masai)	*Conyza bonariensis, Gutenbergia cordifolia*
Enyaru-olmuaate	(Masai)	*Amaranthus hybridus*
Eosin-eiken	(Turkana)	*Alternanthera pungens*
Esegi-enkop	(Masai)	*Tridax procumbens*
Esubukiyai	(Masai)	*Abutilon mauritianum*
Esuguru	(Ateso)	*Tribulus terrestris*
Etaija	(Ankole)	*Commelina benghalensis*
Fuka	(Digo)	*Conyza bonariensis*
Gathenge	(Kikuyu)	*Ageratum conyzoides*
Gathumba	(Kikuyu)	*Datura stramonium*
Gatumia	(Kikuyu)	*Portulaca oleracea*

Gitegenye	(Kikuyu)	*Cyathula cylindrica,*
		C. polycephala
Hangasimu	(Shambaa)	*Chenopodium murale*
Hoko	(Shambaa)	*Phytolacca dodecandra*
Ikengere	(Chagga)	*Commelina benghalensis*
Ilula	(Nyamwezi)	*Acacia drepanolobium*
Inagu	(Kikuyu)	*Solanum nigrum*
Jiji	(Sukuma)	*Rhynchelytrum repens*
Kafumba	(Ankole)	*Galinsoga parviflora*
Kahomba	(Kamba)	*Chenopodium pumilio*
Kakodongo	(Shambaa)	*Cynodon nlemfuensis*
Kakovu	(Luganda)	*Sonchus oleraceus*
Kang'ei	(Kamba,	
	Kikuyu)	*Galinsoga parviflora*
Kasibante	(Luganda)	*Eleusine indica*
Katatula	(Nyamwezi)	*Acacia brevispica*
Katet	(Kipsigis)	*Dichrostachys cinerea*
Kayongo	(Dhopadhola)	*Striga hermonthica*
Kerundut	(Kipsigis)	*Sida rhombifolia. S. ovata*
Kiamata	(Kamba)	*Setaria verticillata*
Kibirosit	(Kipsigis)	*Chenopodium ambrosioides,*
		C. procerum, C. schraderanum
Kibobetyet	(Kipsigis)	*Cynoglossum coeruleum*
Kichoma mguu	(Swahili)	*Bidens pilosa*
Kidadeish	(Shambaa)	*Indigofera spicata*
Kifagio	(Swahili)	*Sida acuta*
Kihumpu	(Shambaa)	*Mucuna pruriens*
Kikatu	(Kikuyu)	*Cyperus rotundus*
Kileleshwa	(Kikuyu)	*Tarchonanthus camphoratus*
Kimavi cha kuku	(Swahili)	*Ageratum conyzoides*
Kitadali	(Swahili)	*Euphorbia hirta*
Kitengejja	(Luganda)	*Pistia stratiotes*
Kituha	(Sukuma)	*Striga hermonthica*
Kivumbavumba	(Luganda)	*Chenopodium ambrosiodes*
Kiwere	(Luganda)	*Rumex abyssinicus*
Kuengi ungi	(Rufiji)	*Pistia stratiotes*

Kwamamaiyet	(Kipsigis)	*Senecio moorei*
Limi ja ngombe	(Shambaa)	*Senecio discifolius*
Lindadongo	(Shambaa)	*Pennisetum clandestinum*
Lubeera	(Luganda)	*Hibiscus cannabinus*
Lugowi	(Swahili)	*Cynodon nlemfuensis*
Lumbugu	(Luganda)	*Digitaria scalarum*
Lumbugu sogule	(Sukuma)	*Digitaria scalarum*
Lusaga	(Kakamega)	*Cleome monophylla*
Lusanke	(Luganda)	*Imperata cylindrica*
Luwugula	(Luganda)	*Triumfetta rhomboidea*
Machacha	(Voi)	*Trichodesma zeylanicum*
Magundulu	(Sukuma)	*Trichodesma zeylanicum*
Mahiu	(Kikuyu)	*Sonchus oleraceus*
Malamata	(Swahili)	*Setaria verticillata*
Malulu	(Swahili)	*Eleusine indica*
Manga	(Chifipa)	*Crotalaria polysperma*
Mangova	(Chagga)	*Chenopodium opulifolium*
Masinde	(Swahili)	*Cyperus rotundus*
Mbalibali	(Swahili)	*Acacia drepanolobium*
Mbariki	(Kikuyu)	*Ricinus communis*
Mbaruti	(Swahili)	*Argemone mexicana*
Mbaya	(Shambaa, (Swahili)	*Rottboellia cochinchinensis*
Mbigiri	(Swahili)	*Oxygonum sinuatum, Tribulus terrestris*
Mbiha	(Swahili)	*Abutilon mauritianum*
Mboge	(Luganda)	*Amaranthus graecizans, A. lividus*
Mchicha	(Swahili)	*Gynandropsis gynandra*
Mchokochole	(Swahili)	*Sida cordifolia, Triumfetta rhomboidea*
Mfunga nyumba	(Swahili)	*Dichrostachys cinerea*
Mhoko, Muhoko	(Kikuyu)	*Phytolacca dodecandra*
Mjimbi	(Swahili)	*Pteridium aquilinum*
Mkalambati	(Swahili)	*Tarchonanthus camphoratus*
Mkanganga	(Swahili)	*Hibiscus cannabinus*
Mkengeta	(Swahili)	*Psiadia punctulata*
Mkumajalaga	(Yao)	*Argemone mexicana*

Mjungwina	(Shambaa)	*Lantana camara*
Mkuwa usiku	(Swahili)	*Leonotis nepetifolia*
Mnavu	(Mbulu, Shambaa, Swahili)	*Solanum nigrum*
Moga wagori	(Nyamwezi)	*Chenopodium album*
Mososoyiah	(Kamba)	*Ageratum conyzoides*
Mpambake	(Swahili)	*Lippia javanica*
Mpungate	(Swahili)	*Opuntia* spp.
Mpupu	(Swahili)	*Mucuna pruriens*
Mpurule	(Swahili)	*Chenopodium* spp.
Mrumbu	(Swahili)	*Typha* spp.
Msekeseke	(Swahili)	*Galinsoga parviflora*
Mshunga	(Pare, Shambaa)	*Launaea cornuta*
Mtimbi	(Swahili)	*Imperata cylindrica*
Mtula	(Shambaa, Swahili)	*Solanum incanum*
Mtunguja mwitu	(Swahili)	*Solanum incanum*
Muana	(Swahili)	*Datura stramonium*
Mubare	(Kikuyu)	*Euphorbia hirta*
Muceege	(Kikuyu)	*Bidens pilosa*
Muhurathi	(Kikuyu)	*Opuntia* spp.
Mukakati	(Swahili)	*Rumex abyssinicus*
Mukasa	(Luganda)	*Senecio discifolius*
Mukengeria	(Kikuyu)	*Commelina benghalensis*
Mukengesya	(Kamba)	*Commelina benghalensis*
Mukenia	(Kikuyu)	*Lantana camara*
Mukigi	(Kikuyu)	*Conyza bonariensis*
Mukinyei	(Kikuyu)	*Euclea divinorum*
Mukunguni	(Kikuyu)	*Chenopodium procerum*
Mukuswi	(Kamba)	*Acacia brevispica*
Mulaa	(Kamba)	*Senecio discifolius*
Mulolo	(Kamba)	*Leucas martinicensis*
Munga	(Kamba)	*Acacia drepanolobium*
Munokelo	(Acholi)	*Tridax procumbens*
Muraa	(Kikuyu)	*Acacia hockii*
Murashe	(Ankole)	*Bidens pilosa*
Murindunguri	(Kikuyu)	*Triumfetta rhomboidea*
Murundu	(Kikuyu)	*Sida rhombifolia*
Mushaga	(Bukoba)	*Cleome monophylla*

Mutengesa	(Luganda)	*Rhynchelytrum repens*
Mutongu	(Kikuyu)	*Solanum incanum*
Muvangi	(Kamba)	*Tagetes minuta*
Muwanika	(Luganda)	*Dichrostachys cinerea*
Muzireti	(Kikuyu)	*Lippia javanica*
Mwamba nyama	(Digo, Giriama, Kibarani)	*Rottboellia cochinchinensis*
Mwengajini	(Swahili)	*Chenopodium ambrosioides*
Ndago	(Swahili)	*Cyperus rotundus*
Ndothua	(Kikuyu)	*Typha* spp.
Ngwata	(Kamba)	*Datura stramonium*
Nkalitongo	(Shambaa)	*Tridax procumbens*
Nnantooke	(Luganda)	*Typha* spp.
Nona nyonya	(Swahili)	*Ricinus communis*
Nsikizi	(Luganda)	*Euclea divinorum*
Ntuuwa	(Shambaa)	*Cyathula* spp.
Nyambo njogabagole	(Sukuma)	*Chenopodium album*
Nyamrungru	(Luo)	*Rottboellia cochinchinensis*
Nyarwanda	(Ankole, Lutoro)	*Portulaca* spp.
Ochok	(Luo)	*Solanum incanum*
Odagwa	(Luo)	*Ricinus communis*
Ogundu	(Luo)	*Sida cuneifolia*
Okuro	(Luo)	*Alternanthera pungens, Tribulus terrestris*
Olenge	(Luo)	*Cymbopogon nardus*
Olgigoi-losirkon	**(Masai)**	***Carduus kikuyorum***
Olgirigiri	**(Masai)**	***Acacia brevispica***
Ol'leleshwa	**(Masai)**	***Tarchonanthus camphoratus***
Olmagutian	(Masai)	*Eleusine jaegeri*
Olmerumuri	(Masai)	*Dichrostachys cinerea*
Ol'momoit	(Masai)	*Solanum nigrum*
Olobobo	(Masai)	*Pennisetum clandestinum*
Oloirepirep	(Masai)	*Cynoglossum coeruleum*
Olonini	(Masai)	*Sida cuneifolia*
Olontwalan	(Masai)	*Crotalaria polysperma*
Olosiro	(Masai)	*Pteridium aquilinum*
Olperesi	(Masai)	*Cymbopogon nardus*
Ol-tagoyie	(Masai)	*Cyathula* spp.

Ol'ungu	(Masai)	*Imperata cylindrica*
Omari	(Luo)	*Chenopodium opulifolium*
Omboga	(Luo)	*Amaranthus* spp.
Ombugu	(Luo)	*Digitaria scalarum*
Omuhoko	(Ankole, Bukoba)	*Phytolacca dodecandra*
Omuhuuki	(Ankole)	*Lantana camara*
Omukurura	(Ankole)	*Achyranthes aspera*
Omusandusi	(Ankole)	*Sida alba*
Omusoga	(Ankole)	*Ricinus communis*
Omuteetera	(Ankole)	*Indigofera spicata*
Ondhong'	(Luo)	*Typha* spp.
Onyiego	(Luo)	*Bidens pilosa*
Orutaratari	(Ankole)	*Eleusine indica*
Osinoni	(Masai)	*Lippia* spp.
Owi pap	(Luo)	*Hibiscus cannabinus*
Oyungu	(Luo)	*Pistia stratiotes*
Ruchwamba	(Ankole)	*Cynodon nlemfuensis*
Rugoli	(Sukuma)	*Cynodon nlemfuensis*
Rutege	(Ankole)	*Cymbopogon caesius*
Sake	(Kamba)	*Gynandropsis gynandra*
Sangari	(Swahili)	*Digitaria scalarum*
Sanguru	(Chagga)	*Digitaria scalarum*
Sasa mlanda	(Swahili)	*Trichodesma zeylanicum*
Segutyet	(Kipsigis)	*Eleusine jaegeri*
Sengesha	(Shambaa)	*Leucas martinicensis*
Shelukungu	(Shambaa)	*Rumex abyssinicus*
Shunga pwapwa	(Shambaa, Swahili)	*Sonchus oleraceus*
Sietera	(Tongwe)	*Acanthospermum hispidum*
Singumbe	(Swahili)	*Cymbopogon caesius, Rhynchelytrum repens*
Sinoni	(Chagga)	*Lippia javanica*
Siratet	(Kipsigis)	*Urtica massaica*
Spirafinca	(Chagga)	*Gutenbergia cordifolia*
Takweiyot	(Kipsigis)	*Carduus kikuyorum*
Talamang	(Karamojong)	*Triumfetta flavescens*
Terere	(Kikuyu)	*Amaranthus* spp.

Thabai	(Kikuyu)	*Urtica massaica*
Tik-jodongo	(Luo)	*Cynoglossum coeruleum*
Tufia	(Shambaa)	*Urtica massaica*
Turura	(Bondei, Shambaa, Swahili)	*Achyranthes aspera*
Ubuti	(Kikuyu)	*Gutenbergia cordifolia*
Uchachai	(Chagga)	*Crotalaria polysperma*
Upupa	(Kiha)	*Mucuna pruriens*
Usuet, Uswa	(Kamasia, Lumbwa, Sebei)	*Euclea divinorum*
Watema	(Kikuyu)	*Pennisetum clandestinum*
Woa	(Kamba)	*Amaranthus thunbergii*

Index

Names in italics are synonyms.
Figures in heavy type refer to the species illustrated.

288